# Build Your Own
# Electric Bicycle

## TAB Green Guru Guides

*Do-It-Yourself Home Energy Audits: 140 Simple Solutions to Lower Energy Costs, Increase Your Home's Efficiency, and Save the Environment* by David S. Findley

*Renewable Energies for Your Home: Real-World Solutions for Green Conversions* by Russel Gehrke

*Build Your Own Plug-In Hybrid Electric Vehicle* by Seth Leitman

*Build Your Own Electric Bicycle* by Matthew Slinn

*Build Your Own Electric Motorcycle* by Carl Vogel

# Build Your Own Electric Bicycle

Matthew Slinn

New York   Chicago   San Francisco
Lisbon   London   Madrid   Mexico City
Milan   New Delhi   San Juan
Seoul   Singapore   Sydney   Toronto

The **McGraw·Hill** Companies

**Library of Congress Cataloging-in-Publication Data**

Slinn, Matthew.
   Build your own electric bicycle / Matthew Slinn.
      p.   cm.
   ISBN 978-0-07-160621-9 (alk. paper)
   1. Electric bicycles. 2. Home-built motorcycles. I. Title.
   TL437.5.E44S595   2010
   629.227'2—dc22                                        2010009182

McGraw-Hill books are available at special quantity discounts to use as premiums and sales promotions, or for use in corporate training programs. To contact a representative, please e-mail us at bulksales@mcgraw-hill.com.

**Build Your Own Electric Bicycle**

1 2 3 4 5 6 7 8 9 0   DOC/DOC   1 6 5 4 3 2 1 0

ISBN 978-0-07-160621-9
MHID 0-07-160621-1

| | |
|---|---|
| **Sponsoring Editor**<br>Judy Bass | **Copy Editor**<br>James Madru |
| **Editorial Supervisor**<br>Stephen M. Smith | **Proofreader**<br>Paul Tyler |
| **Production Supervisor**<br>Richard C. Ruzycka | **Indexer**<br>Judy Davis |
| **Acquisitions Coordinator**<br>Michael Mulcahy | **Art Director, Cover**<br>Jeff Weeks |
| **Project Manager**<br>Patricia Wallenburg, TypeWriting | **Composition**<br>TypeWriting |

*This book is dedicated to my girlfriend Heather, who I love.*

## About the Author

Matthew Slinn is an experienced research scientist and process engineer and has worked on several alternative energy transport applications. During his EngD at the University of Birmingham (U.K.), he worked on the biodiesel process and was sponsored by and worked at Green Biodiesel and BHR Biofuels. During this time he bought his first electric bicycle and quickly recognized the great possibilities for this new form of transport. The new skills he developed while fixing and upgrading this bike won him new jobs in  the field of fuel cells and batteries. He worked on the prototype microcab fuel-cell vehicle and at Oxis Energy developing their lithium sulfur batteries. This book is the product of many years of accumulated knowledge.

# Contents

# Preface

Most people start off with electric bicycles by buying a commercially built one. This is good because it gets you riding straightaway. After you get over the initial novelty of the seemingly magical silent propulsion, you begin to think, "Why can't it go faster?" This is where this book comes in. It covers buying the best commercial electric bicycle for you, upgrading it for more power and speed, building your own more-powerful electric bicycle, learning new skills and how things work, repairing your electric bicycle when it breaks down, and adding modifications to improve it. The content is ordered to make it easy for the novice to jump in and get started and then learn by doing. Few tools are needed, and the skills, such as soldering, are easy to learn. This book will bring you to the forefront of electric bicycle technology and inspire you to take on new electric bicycle projects. Once you have read this book, you will want to keep it handy as a manual to guide you through any electric bicycle problems that you encounter or projects that you wish to undertake. Unlike cars, a separate user manual for each make and model of vehicle will not be needed. This book covers all makes of electric bicycles because the technology is simple and the same.

*Matthew Slinn*

# Acknowledgments

I would like to thank the moderators of, and everyone who has contributed to, the Endless-Sphere Technology Forum (www.endless-sphere.com). Without them, this book would not be possible. Forums like this are an open source of ideas, where science and engineering happen at lightening speed. This wealth of knowledge and these new discoveries are a benefit to us all.

# CHAPTER 1

# Introduction

This book is a practical guide to electric bicycles aimed at the novice and intermediate riders of electric bicycles who are interested in learning more and getting the most out of their bicycles. It will be a fun and "hands on" approach that will cover buying, building, riding, upgrading, modifying, and maintaining your electric bicycle. Electric bicycles are far simpler than cars, essentially consisting of three main components (i.e., battery, controller, and motor) and so are much more user serviceable with a little know-how. Whether you use your electric bicycle for everyday commuting, shopping, racing, or just to show off and impress your friends, this book will help you to succeed.

## 1.1  What Are Electric Bicycles?

Electric bicycles are bicycles that have an electric motor and batteries that power the bicycle and assist with pedaling. An electric bicycle is a hybrid of electric and pedal power. Electric bicycles are not motor bikes. They are usually limited by law to a specific power so that they still qualify as bicycles and are exempt from registration fees, insurance, needing a driving license, and the department of safety regulations. The specifications that electric bicycles must meet to be exempt from motor vehicle registration vary from country to country and state to state. In Europe, the current limit is 250 W maximum continuous power and 40 kg/88 lb maximum weight with no limit on peak power. In the United States, the limits vary from state to state, with some states allowing 750 W and some not allowing electric bicycles at all. In the West, electric bicycles are new technology, and most people don't know anything about them. In the Far East, however, electric bicycles are big business. In China, electric bicycles outnumber cars by four to one.

You may find that your electric bicycle draws a lot of attention. When parked, you will see people stare and try and figure out what all the strange-looking bits are for. If you are seen with an electric bicycle, you might get asked lots of questions by inquisitive people. The most common question people ask is: "Does pedaling charge the battery?" The answer is no. You use up the battery when riding and charge it from the socket when you get home. You can pedal to go faster, but the idea of having an electric bicycle is to pedal less. If the battery were to be charged by pedaling, then you would have to do just as much pedaling as a regular bicycle. This question probably comes from confusion with hybrid cars, where the engine charges the battery. Some of the most frequently asked questions are answered in Table 1.1.

**TABLE 1.1**  Top 10 Most Frequently Asked Questions about Electric Bicycles

| | | |
|---|---|---|
| 1. | Does pedaling charge the battery? | No. You use up the battery when riding and charge it from the socket when you get home. |
| 2. | How often do you have to charge the battery? | Plug in when you get home after every ride. That way you never have to worry about running out of power. Charging is easy, just like charging a laptop or a mobile phone. |
| 3. | Is that a car battery? | Lots of different batteries are used on electric bicycles, but car batteries surprisingly are the worst choice by far. |
| 4. | How far does it go? | Most commercial electric bicycles are designed to go only 10 to 20 miles, but much more is possible if you have more or bigger batteries. |
| 5. | How fast does it go? | Most commercial electric bicycles go 15 to 20 mph, but much more is possible. |
| 6. | How much does it cost? | Most commercial electric bicycles cost between £500/$750 and £1,000/$1,500. |
| 7. | How heavy is it? | Most commercial electric bicycles weigh between 25 kg/55 lb and 40 kg/88 lb. |
| 8. | How long does the battery last? | This depends on the type of bicycle, but most last for several years. |
| 9. | That is cheating! Why don't you just pedal? | Because I like to go fast and not sweat. |
| 10. | Why don't you put a wind turbine on the front? Then you could generate energy to charge the batteries. | Because I don't believe in perpetual motion! |

Most electric bicycles don't require you to pedal, and some custom electric bicycles don't even have pedals at all. Some electric bicycles have a governor-type mechanism that requires the rider to pedal before power is given. These are called *pedelecs*. On some of the cheaper commercial electric bicycles, you need to pedal a little because the batteries are so weak. On the more powerful electric bicycles, most people choose to pedal slightly because it's a fun "hands on" activity to do. Pedals are like instant feedback, telling you how fast you are going. Accelerating so fast that you no longer need your lower gears is a wonderful feeling. Then, on some electric bicycles, there is the really scary feeling of going so fast that pedaling makes no difference; you can't physically keep up with the motor, even in your highest gear. This feedback tells you that you need to slow down! Pedaling also can be useful if you get caught out and run out of juice or if you have eaten too much over Christmas and decide that you want to lose a few pounds. With electric bicycles, you can decide how much exercise you want to do!

## 1.2  Why Electric Bicycles?

### 1.2.1  Freedom and Convenience

An electric bicycle will take you door to door without breaking a sweat. It has the speed advantages of a motorbicycle and the freedom of a bicycle combined. An electric bicycle can be parked anywhere, is cheap to run, and is fast and fun. An electric bicycle is much more practical than a regular bicycle because you can carry more stuff and it's safer. Some die-hard cyclists will frown on your electric bicycle, dismissing it by saying that it is too heavy or that it's not environmentally friendly. These people are misguided, old fashioned, and *wrong*. Carbon dioxide ($CO_2$) emissions for electric power production are lower than for human power (Lemire-Elmore, 2004). Weight is also irrelevant when it is what powers the vehicle. Not everyone can be a muscle-bound, sweaty pedal pusher, and not everyone wants to be. Even the most purist cyclist will agree that there's no way you can carry a week's worth of shopping on a normal push-bicycle and still cycle at a sensible speed (Figure 1.1). Having the convenience of an electric bicycle even might allow you to totally replace your car use, meaning that you could save thousands on insurance, fuel, and taxes and save the planet, too.

**FIGURE 1.1**   A week's worth of shopping can be carried easily on an electric bicycle in either a backpack or pannier bags. The motor will power the bicycle without you having to sweat. Now there is no need to have a car just for shopping.

### 1.2.2  Safer Than Regular Bicycles

With an electric bicycle, you have a big power source for safety features such as powerful lights and horns that you just don't have on regular bicycles. Because you're not exhausted from pedaling all the time, you can sit back, relax, and get on with the important things—such as anticipating traffic movements and avoiding hazards. You also can keep up with traffic instead of having it whiz by you at the side of the road. This makes cycling feel much safer, although there are no statistic on whether it actually is safer. Keeping up with traffic means that fewer cars have to overtake you, and this protects you from rear-end collisions. Keeping up with traffic also means that drivers have a longer time in which to notice you when approaching you from behind. This, in combination with lights that are 20 times more powerful, should make electric bicycles much easier to spot than regular bicycles. Collisions owing to people pulling out in front of you may be worsened compared with regular bicycles, but to some extent you have control over what's in front of you and can anticipate with braking. Electric bicycles are a lot slower than motorcycles, so accidents will be easier to walk away from. With an electric bicycle, you feel more equal to other road users, and you don't feel like a second-class road user, as you do when riding a regular bicycle. This may give riders of electric bicycles a more polite, relaxed attitude toward other road users. Riders of electric bicycles will be more likely to stop at red lights and pedestrian crossings and slow down around hazards. This is so because electric bicycles have a battery pack that will boost riders back up to speed, and they know they don't have to sweat it out.

Road safety in general is improved if people switch to lightweight electric bicycles instead of cars. In collisions with pedestrians or other cyclists, bicycles will do much less damage than cars. Car drivers feel too protected inside their cars and take more risks as a result. Bicycle riders will be more cautious, and there will be fewer accidents as a result. On a bicycle, it's difficult to get distracted by a mobile phone, cigarette, or applying your makeup. The increased visibility that bicycle riders have also makes for safer roads than in our car-dominated society.

The first time a person rides an electric bicycle, he or she is left smiling ear to ear by a phenomenon that has been called the *electric vehicle grin*—a result of the feeling of movement without any effort or noise. For many of us, the grin does not go away as long as our batteries are good and our bicycle doesn't let us down.

### 1.2.3  The Benefits of an Electric Bicycle

There are many people who could benefit from owning an electric bicycle. The cost of car ownership is now prohibitively high for many people, especially the young, who get hammered by the cost of insurance. Using an electric bicycle instead of a car to get around can save you thousands of dollars a year and is often a much faster mode of transport in congested cities. With a bicycle, you can filter through traffic to the front of the line, and you can take shortcuts where cars are prohibited. Speed cameras are causing record numbers of people to lose their licenses, and without a license, an electric bicycle is the fastest thing you can drive. Parking problems, traffic jams, and road congestion are making many urban car journeys unfeasible. Years ago, a car brought a feeling of freedom. Now motorists are easy targets for speed traps and parking meter attendants. Buses are often unreliable, can be expensive, and are universally disliked. Cycling is a good option in cities, but many people don't like to sweat all the time. Electric bicycling is the answer!

### 1.2.4  Cost Savings

The biggest cost saving of having an electric bicycle is that you don't need a car, which saves you the cost of gas, insurance, maintenance, taxes, depreciation, and parking fees and tickets. To replace car use completely, we

FIGURE 1.2    Joined-up transportation. The bicycle and train are an amazing combination that will take you long distances quicker and cheaper than any car. The electric bicycle makes this transport choice even better.

need a solution for long-distance travel. That solution is to use a bicycle in combination with a train. The bicycle and train are an amazing combination that will take you long distances quicker and cheaper than any car (Figure 1.2). The electric bicycle makes this transport choice even better. We can compare the cost of the electric bicycle–train combination versus private car ownership. Every situation is different, but I would think that it's difficult to find a case where the electric bicycle–train combination is not as cost-effective as private car ownership. Table 1.2 shows my annual cost estimation for my 40-mile daily commuter journey, 24 weekend trips per year, and taxis for going out at weekends.

For the scenario in the table, the capital cost of buying the new electric bicycle is at least half as much as buying a second-hand car, and on top of that, you would save £2,300/$3,450 a year on operating costs. If you live and work in a city, then the savings would be even better for the electric bicycle because the fixed costs of car ownership (insurance, tax, maintenance, and depreciation) would begin to dwarf the variable distance costs (fuel or train fares). Living and working in the same city means that you can use your electric bicycle more and cut out most of the train fares. Season's ticket train fares seem to be priced very close to the equivalent fuel costs for the same journey done by car. Savings could be increased further if you were to use the bicycle instead of taxis to go out in the evenings where it is impossible to park in the city. However, since this might not be practical if you

TABLE 1.2   Cost Comparison Between Car Ownership and Electric Bicycle Ownership for My 40-Mile Daily Commute, 24 Weekend Trips per Year, and Taxis on Weekends

| Electric Bicycle and Train Combination | | Car | |
|---|---|---|---|
| *Capital cost* | | | |
| Bicycle (including battery) | £500–1,500/ $750–2,250 | Car | £1,000–2,000/ $1,500–3,000 |
| *Annual operating costs* | | | |
| Electricity | £10/$15 | Gas | £2,500/$3,750 |
| Train fares | £2,226/$3,339 | Train fares | £0/$0 |
| Insurance (third party) | £35/$52 (optional) | Insurance (third party) | £700/$1,050 |
| Maintenance (new batteries, etc.) | £200/$300 | Maintenance | £600/$900 |
| Depreciation | £0/$0 | Depreciation | £500/$750 |
| Tax | £0/$0 | Tax | £125/$187 |
| MOT | £0/$0 | MOT | £40/$60 |
| Taxis (for going out in the evening) | £300/$450 | Taxis (for going out in the evening) | £300/$450 |
| Parking tickets and fines | £0/$0 | Parking tickets and fines | £300/$450 |
| *Total annual operating costs* | | | |
| Bicycle and train combination | £2,771/$4,156 | Car | £5,065/$7,597 |

have to wear a nice suit or dress, I have assumed that riders of electric bicycles and car drivers spend the same amount on taxis annually. The cost of insurance also could vary considerably. For example, for a 17-year-old male driver, the cost of third-party car insurance is currently £4,500/$6,750 irrespective of which car he might drive (www.confused.com, 2009). One of the only reasons one might "need" a car is to impress members of the opposite sex. I suspect that many people still have car ownership as one of their "standards" for potential suitors!

An electric bicycle will use about 20 Wh/mile, and electricity costs about 11 pence (16 cents)/kWh. Therefore, the energy cost for an electric bicycle is about 0.2 pence (0.3 cents)/mile, which is so low that it's almost insignificant compared with other costs. Battery replacement is the main cost for electric bicycles because batteries are really a consumable item. Most good batteries sold for electric bicycles are rated for 1,000 discharge cycles and cost around 50 pence (75 cents)/Wh capacity. Therefore, we can calculate that the operating cost of battery replacement is around 1 pence (1.5 cents)/mile (if you look after your battery). In Britain, fuel costs about £1.06/liter or $6/gallon, which is mostly tax, and the average car in Britain does about 8 miles/liter (Mackay, 2009). Therefore, car fuel costs about 13 pence (20 cents)/mile, which is over 60 times the energy cost of an electric bicycle. When you include the operating cost of battery replacement, the electric bicycle is still 13 times cheaper to run than a car. A small gas scooter will get around 27 miles/liter or 100 miles/gallon, which works out to be around 4 pence (6 cents)/mile. This is four times the operating cost of an electric bicycle. Since oil is a scarce resource that is fast running out, we can expect the price of fuel to continue to increase in the future, so the savings of electric bicycles can only improve.

### 1.2.5  Time Savings

Car journeys are often slow owing to traffic congestion. In cities, an electric bicycle usually will be faster because it can slip to the head of traffic lines. When traveling longer distances, trains (or planes) are faster and more direct and, in conjunction with your electric bicycle, could result in a considerable time savings. In addition, on the train, you are free to do other things than concentrate on driving. Much of this book was written while I was on a train. If you value your free time, then 60 free minutes on a train per day at £20/$30 per hour is worth £5,000/$7,500 each year. One could do a similar calculation for time saved by riding an electric bicycle rather than a regular bicycle.

### 1.2.6  Saving the Environment

The world is running out of oil, and with it, the price of all forms of energy will get much more expensive. This will make cars, electric or otherwise, prohibitively expensive for most people. In the future, it's thought that global warming will be a serious threat to human progress and prosperity. Global warming is caused by carbon emissions, which are brought about, among other things, by the burning of fossil fuels—our main source of energy. The electric bicycle often can replace the use of a car and do the job much more efficiently. A liter/gallon of gas contains about 10 kWh of energy; therefore, the average car in Britain/United States uses about 1,250 Wh/mile, or 60 times as much energy as an electric bicycle. This means that an electric bicycle can go at least 60 times farther than a car on the same amount of energy.

There are some inefficiencies in the production and distribution of electricity. Production of electricity is about 30 to 45 percent efficient. However, even with these numbers, there is still a massive reduction in emissions and energy use when an electric bicycle is chosen over a car. Mackay (2009) addresses the math behind energy use and energy policy. He calculated a 100 times reduction in energy use (at the point of delivery) for a bicycle compared with a car. Therefore, the amount of energy that an electric bicycle uses is insignificant compared with the energy use of a car and so low that it easily could be generated by a renewable energy source such as solar power, whereas it would not be practical to power a car with renewable energy—the cost and land area required to do so would be beyond the means of most people.

First reactions to electric bicycles can be quite interesting. People generally don't like change, and some of them initially argue that electric bicycles are "just wrong." Such a negative reaction may come from car drivers who feel guilty about their pollution when faced with a new alternative or from cyclists who think they are morally superior for all the hard work they do. These people are in the minority, and fortunately, most people are just very interested and even may end up buying an electric bicycle of their own. Some of the negative reactions derive from people's assumptions about what is environmentally friendly and what's not.

Many people assume that putting a battery and motor on a bicycle will only increase its energy use, but this is not true. In an electric bicycle, the motor replaces human power, which is achieved at the cost of eating more food. Lemire-Elmore (2004) calculated the total life-cycle energy use of an electric bicycle with different battery technologies and compared the results with energy consumption by a pedal cyclist. Taking into account factors such as emissions from production, delivery, and replacement of batteries, he concluded that there was up to five times fewer emissions from electric bicycles than from human-powered bicycles where the rider has an average Western diet. This is not surprising when you consider all the energy use and emissions from the food chain, including farming, transport, manufacture, refrigeration, cooking, and the 25% efficiency of the human body for converting food into energy. If you have a 5-minute shower (3 kW) after a sweaty pedaling session, then this will use more energy than cycling 12 miles on an electric bicycle. Human power is not as green as people assume. The electric bicycle is probably the most efficient practical mode of transport in the world.

### 1.2.7 The Future of Transport

Cars may become slightly more efficient in the future. They may have new types of engines, batteries, or even fuel cells, but a car can never be as efficient as a bicycle. The reason lies in the concepts of drag and weight. Cars are designed big and heavy with the capacity to carry 4 to 5 people, but most cars are opperated with only a single occupant. This "misuse" of the vehicle is what causes the inefficiency. If an efficient car is opperated at 100% occupancy then the energy (per person, per mile) starts to aproach that of an electric bicycle. The large frontal area of a car and its weight mean that it will always require a large amount of energy to move. This energy demand does not change if you change the technology powering the car without also changing its weight and/or aerodynamics. The only way to improve a singly occupied car's fuel consumption significantly is to make it lightweight and torpedo shaped, in effect making it more like a motorbike. Car manufacturers know this and have made many fuel-efficient concept cars, but these cars never get built because of worry about consumer acceptance of "weird looking" cars. As the cost of fuel becomes more of an issue, aerodynamic cars will gain in popularity. Having said this, the gas engine is fairly inefficient, and there are small gains to be made by changing engine technology.

Don't believe the hype about hydrogen fuel cell cars or electric cars saving the world. Hydrogen and electricity are just energy carriers and not energy sources like oil, coal, solar, or wind energy. Hydrogen and electricity have to be made from our limited energy sources. When all the energy comes from fossil fuels, then these advanced vehicles are often just as costly and polluting as regular vehicles. Several scientific studies show that future cars with fuel cell power-trains will produce about the same emissions as gasoline vehicles today (van Mierlo, Maggettoa, and Latairea, 2006). This is so because all the inefficiencies of the energy conversion stages involved in making, storing, and using hydrogen add up to the same inefficiency as using gasoline in an engine. Battery electric cars fare a bit better and have the potential to roughly halve the greenhouse gas emissions of regular cars (van Mierlo, Maggettoa and Latairea, 2006). Hybrid cars are also good because the regenerative braking re-aptures wasted energy. Plug-in hybrids are a good compromise because they add the range of a gas engine to the efficiency of electric drive. The efficiency of electric vehicles is linked to the power station efficiency. This may improve in the future if we go renewables or nuclear, but it also could get worse as oil runs out and nations start switching to coal for their power.

## 1.3 Safety with Electric Bicycles

**Warning:** Working on electric bicycles can be dangerous! There are many possible hazards, including battery fires and explosions, electrocution, crushing, tripping, and falling. Assess the risks before working on your electric bicycle, and wear the appropriate safety and personal protective equipment (PPE), for example, goggles, gloves, protective shoes, and thick clothing.

### 1.3.1 Electrical Safety

Do not touch live battery terminals. Anything over 48 V is considered a high-voltage source and could result in a lethal electric shock. However, it's difficult to get an electric shock accidentally from an electric bicycle battery because dry, intact skin is a good insulator. Main circuit box voltages are higher and much more dangerous, however, and shouldn't be messed with. Do not work on anything that is connected to the main circuit box; always disconnect it first! Do not perform maintenance on the battery or the bicycle while it is being charged. Do not connect the charger to the main circuit in the rain or when you or your bicycle is wet. Do not charge your bicycle in the rain. A fuse will not protect you from an electric shock; only a correctly installed circuit breaker will. Everything should be grounded correctly for the circuit breaker to work. I cannot be held responsible for any accident, injury, or damage to property caused by your work following the instructions in this book.

### 1.3.2 Battery Chemical Safety

Be very careful when working on batteries. The batteries on electric bicycles contain a lot of energy and can be very dangerous. Some batteries are unstable and prone to fire if mistreated. On rare occasions, some batteries may contain manufacturing defects that cause them to combust spontaneously for no reason. Poor quality control during manufacture is usually to blame for these faults, and has been the cause of several battery recalls in laptops and other consumer gadgets. Lithium polymer (Lipo), lithium colbalt (LiCo) and lithium-ion batteries have unstable chemistries and will catch fire or explode if overcharged, short-circuited, or physically damaged. If you don't believe me, watch the videos on youtube.com! If you are new to electronics or electric bicycles, you are advised not to use these batteries. In fact, most people avoid using these batteries altogether owing to the potential risk of fire. Treat these batteries with extreme care. Only charge the battery outside your house in a nonflammable location in case you have a charger malfunction that results in a battery explosion. Other battery chemistries generally are safer, and some have built-in safeguards that prevent dangerous failure. The fire risk of batteries is one of the main things holding back electric cars. Large electric vehicle (EV) battery packs often contain hundreds of cells, which increases the chance of failure. The high temperatures caused by some cell failures can cause other neighboring cells to fail, which, in theory, could cause a chain reaction that would destroy the EV.

The chemicals in batteries are often toxic. If you overcharge a battery and it vents electrolyte, avoid breathing the fumes. If there is a battery fire, do not breathe the toxic fumes. Get away as fast as possible, and call the fire department if necessary. Disconnect any smoking battery from the charger if it is plugged in, but only if safe to do so. If the battery is smoking, put it outside if it is safe to do so. If the battery is in a safe location outside, then just let it burn. If the battery is on fire inside, then extinguish the fire with a fire blanket and a $CO_2$ extinguisher, but only if it is safe to do so. It may not be possible to extinguish the fire because of the chemicals involved. Do not use water fire extinguishers because lithium reacts with water. Wear protective equipment when extinguishing a battery fire in case the battery explodes or bits are blown around by the fire extinguisher. The highest risk is when you are working on an open battery. When working on batteries, avoid using metal tools whenever possible; work on a clean, empty, nonconductive surface, and don't wear jewelry. If a cell is broken or damaged, do not attempt to fix it. It is impossible anyway because individual cells are not made to be user serviceable, and building them is a very intricate and complicated process. Having said all this, batteries, if they are looked after, are usually very safe. Dangerous battery incidents are rare.

### 1.3.3  Road Safety

When riding an electric bicycle, always wear a helmet; this is the law in some states and countries. In addition, it is advisable to wear thick protective motorcycle gloves, strong shoes, thick trousers (denim jeans are okay), and a jacket. Electric bicycles are not as fast as motorcycles so the protective equipment doesn't have to be as cumbersome. However, you are not invincible, and if you fall off, it will hurt, and you may sustain an injury. Your hands are usually the first thing to touch the concrete in an accident, so thick motorcycle gloves are very helpful, and many cyclists wear them. Make sure that your brakes are good enough to stop you in two car lengths at 20 mph. If the road is wet, watch out for downhill sections where braking is needed for a turn into a side road. You will skid dangerously if you try to brake too quickly.

Always obey the traffic and vehicle laws in your country. Follow the highway code as it applies to cyclists. Follow any special electric bicycle laws in your country. Don't speed or show off. Ride safely, and share the road with other users. It's a good idea to take a motorcycle driving test or cycling proficiency test. If you are new to cycling, then some training will greatly enhance your confidence and keep you safe. Most accidents happen to inexperienced cyclists. Use lights on your bicycle at night; this is the law in many countries. Consider wearing bright florescent material on your jacket, bicycle, or your rucksack. Cycle a meter/yard or so from the curb. This will make you more visible to motorists, and they will not try to squeeze past dangerously. This also will protect you from getting "doored" by people getting out of parked cars. Take the appropriate lane at junctions, and behave like a car. If you are timid at junctions, other drivers will not understand your intention. Many accidents happen with large vehicles such as trucks or buses. These vehicles have large blind spots and large turning circles. Don't try to slide past these vehicles, especially on the inside, at traffic lights because they will not see you and could squash you against railings, parked cars, etc. Don't cycle on the sidewalk because it is more dangerous than cycling on the road. Car drivers will not look for vehicles there, and you are likely to get hit while crossing a driveway when a car is turning in.

## 1.4  Legality of Electric Bicycles

In Europe, the legal power limit on electric bicycles is not very clear. The law says, "Maximum continuous rated power output of the motor shall not exceed 250 W." This leaves a massive loophole because there is no limit on peak power. As long as your motor eventually overheats at 250 W continuous power, then one would assume it is okay to have a 2-kW motor. You could complete a short journey pulling a peak power of 2 kW through a 250-W continuous motor, and it would get very hot. Then, when you reach your destination, you just let it cool down, and you still would be within the letter of the law. The law does not specify how power is tested. The efficiency of electric motors varies depending on speed, where they are less efficient at low speeds. The speed at which the measurement is to be taken is not specified. Therefore, one could claim that because the motor overheats at 250 W continuous power when the bicycle is doing only 1 mph, pulling a bus, for example, then this satisfies the criteria. The law specifies output power, which is very difficult to measure. To measure maximum continuous rated power output, you would need a dedicated laboratory equipped with a rolling road. The measurement would be a destructive test on all motors that obeyed the law because if they were rated below 250 W maximum continuous power, then you would destroy them by putting 250 W continuously through them. The only practical part of the European or UK law is the weight limit. It's easy to do roadside testing of bicycle weight with scales.

Peak power of 250 W is such a small amount of power that it does not allow riders to climb even moderate hills. In the United States, federal law states that electric bicycles can have a "motor not exceeding 750 W power or 20 mph." This is a much better law because it states the peak power, which is more easily measurable, and the top speed, which also is measurable. Peak power of 750 W is a reasonable amount of power that will propel a cyclist up most hills. Peak power of 750 W is close to what an athlete cyclist or a horse would be able to generate, which also falls outside motor vehicle codes. A speed of 20 mph is reasonable because it is similar to what an athlete cyclist would be able to do.

"The electric motor must not be able to propel the machine when it is traveling at more than 15 mph." This part of European law also has caused some confusion because it requires special speed-sensing circuitry that most electric bicycles from China don't have. As a result, some manufacturers just sell electric bicycles that have low-power motors that can't exceed 15 mph on flat ground. These bicycles are completely useless at climbing hills. Some manufacturers sell bicycles that have an off-road switch that delimits the bicycle to faster speeds. Cyclists are allowed to pedal faster than 15 mph so it's not possible to prove failure to comply with this law using a speed camera. If the rider pretends to pedal, then it's impossible for a bystander to tell if the power is provided by the rider or the motor.

Under European Union (EU) law, only power-assisted bicycles meeting the pedelec classification are considered to be pedal cycles. The law stipulates that "the motor must be activated by the rider's pedaling effort, and the power must cut out completely whenever the rider stops pedaling." To comply with this law, a pedal-assist system (PAS) can be implemented in two different ways: cutting off power when the rider stops pedaling or trying to provide power proportional to pedaling effort. PAS is basically a special controller that decides when it wants to supply the power to the motor based on your pedaling. The idea is that it senses your pedaling and braking and supplies power to the motor accordingly, making the bicycle feel like a normal bicycle that is just a little bit faster. The reason it is bad in practice is because it depends on software to interpret your movements and takes control away from the rider. Sometimes the software is bad, and there is a delay between sensing and providing power. This delay can cause the motor to provide full power when you actually want to stop or leave you struggling when the traffic light turns green! It's very difficult to sense pedaling effort because that would require a torque sensor, which is expensive and impractical. Most pedicels have only a pedal speed sensor and try to interpret the rider's power demand from his or her pedaling speed. This is completely useless because if the rider stops when going up a hill, the PAS cuts out power and makes it very difficult for the rider to get started again. PAS makes an electric bicycle feel uncontrolled, unreliable, and dangerous. The PAS system is very counterintuitive for people who are used to bicycles or motor bikes. The politicians who passed this law probably have never used an electric bicycle.

The EU electric bicycle laws create a gray area that blurs the boundaries between low-powered mopeds and bicycles. With the current EU laws, scientific testing would be needed to prove that someone is illegally riding a moped without insurance and not simply riding a bicycle. Police forces don't do forensic tests unless someone is actually hurt or killed. Therefore, as long as riders of electric bicycles stay close to the spirit of the law by riding like cyclists, there are unlikely to be any legal troubles, even with 2-kW motors. If riders of electric bicycles are caught overtaking cars with motor bike–like power, then the classification of the electric bicycle might be called into question. For most riders, reaching motor bike–like power is unlikely considering the state of the technology. Police and lawmakers probably were waiting for lots of accidents to happen, such as when there was a crackdown on go-peds and gas scooters. Since most riders of electric bicycles are law-abiding citizens just trying to commute to work, there has been no spate of accidents, and EU lawmakers are currently looking at ways to increase the power limit on electric bicycles and clarify the law.

# CHAPTER 2

# Your First Electric Bicycle

## 2.1 Commercial Electric Bicycles

There are many different electric bicycle retailers and distributors. Most retailers sell direct on eBay, and in addition, there is a thriving secondhand market on eBay as well. A commercial electric bicycle is a good way to introduce yourself to the world of electric bicycles (Figure 2.1). It will get you all the components at a good price, and of course, the bicycle comes ready assembled. This is a good starting point from which you can learn to upgrade and improve your electric bicycle experience.

If you're interested in buying a new electric bicycle, you can get something fairly good in the price range of £300 to £600 or $450 to $900. For a top-of-the range electric motorbike, you can pay any thing up to £5,000/$7,500. The battery is the most expensive part on an electric bicycle. As a guide, good batteries sell on eBay for around 2 Wh/£ or 1.3 Wh/$. Motors usually cost around £100 to £200/$150 to $300 depending on power, and controllers can cost £50 to £150/$750 to $2,250 depending on power. Aim to buy a bicycle for a price equal to its components' prices. Secondhand electric bicycles can be found on eBay for as little as £200/$300, although they may need some work. It's best to have a bit of experience first before buying secondhand because you won't know what good/bad points to look out for.

When buying on eBay, always check the seller's feedback. If it's a new electric bicycle, make sure that the seller has lots of positive feedback for selling electric bicycles. Positive feedback for selling cheap stuff such as women's clothing doesn't count. Watch a few auctions first to see what the going price is. If there is a retailer near you, go there and try some of the electric bicycles. If you are buying from an Internet retailer, you might want to test the seller's after-sales service by phoning the company and inquiring

**FIGURE 2.1**   A typical commercial electric bicycle.

about buying a second battery or some spare parts. If the people are too busy to deal with you, then buy elsewhere.

### 2.1.1  Buying an Electric Bicycle: The Motor

The two main motor formats on retail electric bicycles are hub motors and chain-drive motors. It's probably best to choose the hub motor, which is used in most electric bicycles, because the motor can be swapped easily to another frame. This will make maintenance easier. You can upgrade any bicycle with a new hub motor. It just requires swapping the wheels (see Section 3.1). Chain-driven electric bicycles should not be ruled out because the frame will have a convenient mounting point for chain-drive motors that offers more upgrading possibilities for the expert electric bicycle rider. It's not an easy task to upgrade a chain-drive motor because you have to equally tension both the pedal and the motor chains. There are limited things to adjust, and the "sweet spot" can be difficult to find.

Two types of motors are available: brushed and brushless. These are explained in detail in Section 4.1.1. Brushed motors are more reliable, simpler, and cheaper, but brushless motors are usually more efficient. There are fewer things that can go wrong with a brushed motor, and they are easier to upgrade or overclock. Commercially available electric bicycles mostly use brushless motors because the increased efficiency makes the small battery last longer. However, models with brushed motors are also available.

Try to get the highest powered motor available because most of the commercially available electric bicycles are quite weak. A difference of just 50 W between two motors is noticeable, especially when climbing hills. Usually, the faster motors are larger in size to accommodate heat dissipation. Some of the old Powabykes are quite good for this reason. I would avoid small geared hub motors because some, such as the Tongsin and the Tarn, have a bad track record for malfunctions.

Legal maximum power requirements vary among countries, and manufacturers tend to meet these requirements just by changing the voltage of the battery to make the bicycle faster or slower. This is quite a blunt approach because it means that motors in lower-power bicycles are the same as those in higher-power bicycles but are underclocked to meet legal requirements. In such situations, a simple overclocking can be achieved by adding more batteries in series. See Section 6.1.5 for a detailed "how to" about overclocking. Some of the better electric bicycles use higher power motors but incorporate speed sensors to limit speed to legal requirements. This is good because these bicycles have much better hill-climbing ability.

### 2.1.2 Buying an Electric Bicycle: The Battery

When comparing batteries, it is the overall capacity that is important and not necessarily the volts or amps. Battery capacity is measured in watthours of energy, where watthours = volts × ampere-hours. The retailer might not state the capacity, so you may have to do this calculation for yourself.

The battery is often the most expensive part of the package, and a good battery can make the difference between a good and a "rubbish" bicycle. Try to go for an electric bicycle with at least a 200-Wh battery pack. Don't believe any retailer's range specification. These are usually ridiculously exaggerated to be three or four times more than actually will be achieved. There are no standard tests on electric bicycles yet, and because range varies depending on how much you contribute by pedaling, retailers basically can claim whatever they want. As a guide, 20 Wh will be needed per mile with no pedaling. Therefore, a bicycle with a 24-V, 7-Ah battery will have a range of 8 miles when new without pedaling. As the battery gets older, this range will decrease. Keep this in mind when buying secondhand, and ask to ride the bicycle before you buy it. More powerful, faster motors will require slightly more watthours per mile.

It may be a good idea to buy a couple of spare battery packs with your electric bicycle to increase the range or for future upgrades. You don't want to have to pedal home on an empty battery. Bicycles with a lithium or nickel battery (e.g., NiMh, NiCd) are much better than lead-acid batteries (SLA). In fact, it is advisable to avoid lead-acid batteries altogether. SLA batteries weigh twice as much as NiMh batteries and three times as much as lithium batteries. SLA batteries also have less usable capacity than lithium or NiMh batteries because of something called the *Peukert effect*. To take account of the Peukert effect for electric bicycle use, you need to divide the rated capacity of an SLA battery by half for comparison to other batteries. Lead batteries also have half the lifetime of nickel or lithium batteries. Because of these poor qualities, SLA batteries are several times cheaper than other batteries. With lead acid batteries you need a 1 kg/2.2 lb of battery per mile traveled. If a retailer's advertisement does not state what battery is on the electric bicycle, then it is probably a lead battery. If you need a folding bicycle for trains, then definitely choose a bicycle with a lithium battery because it will weigh less.

### 2.1.3 Buying an Electric Bicycle: The Bicycle

There are two styles of electric bicycles: those that look like bicycles and those that look like scooters. It is advisable to avoid models that look like scooters to avoid being pulled over by the police for not having a registration plate on what looks like a motorcycle. You then will have to explain that your bike is actually a bicycle in disguise, which is a waste of police time and your time. Not only this, but motorists will expect you to be going a lot faster, which could result in confusion and get you beeped at or even involved in a collision. My preference is for a vehicle that is faster than it looks, so I recommend buying an electric bicycle that looks like a traditional bicycle. Avoid bicycles with a pedal-assist system (PAS) if you can. Only buy electric bicycles with throttle control.

Keep in mind that most of the cheap electric bicycles will steadily fall apart on you. Pedals, mud guards, and racks will fall off; frames will rust quickly; and controllers will brake. With this in mind, try to buy a sturdy-looking bicycle that has standard specification components that can be replaced easily.

The extras you should look for in an electric bicycle are a rear carrier, suspension, disc brakes, stand, a lockable battery, and mud guards. A rear carrier is essential for panniers to carry extra batteries or luggage. It's difficult to fit a rear rack to a dual-suspension bicycle. Suspension is usually a gimmick unless you plan to upgrade later and ride around at 30 mph. The rear suspension pivots from a single point, but because there is usually some play in the bearings, the suspended rear section can wobble sideways. This wobble makes the bicycle unstable at fast speeds or if you place heavy batteries in the rear rack. Mud guards are essential for commuting in all weather. A lockable battery compartment is important because thieves will know that these batteries are worth money. If you need to go on the train or bus with your electric bicycle, you will need to get a folding bicycle. Folding electric bicycles also come in handy if you need to retrieve your car after a night out. Folding electric bicycles usually have smaller wheels and a shorter length, which makes them handle and feel worse to ride than normal bicycles. There can be a bit of a tradeoff between folded size and rideability.

Folding electric bicycles are also difficult to upgrade because there is no space for additional batteries. Front or rear baskets are highly useful for serious commuting purposes, so don't dismiss them on looks. Disc brakes are highly desirable on an electric bicycle because you will be going twice the speed and carrying more weight than with a regular bicycle. Look for bicycles that have disc brakes or disc brake pegs for a cheap upgrade later on. Disc brakes stop you a lot faster, and they require almost no maintenance. Some electric bicycles come with lights powered by the main battery. This is better because the lights will be brighter and will be harder for thieves and vandals to steal.

Most electric bicycles for sale are made in China, and they vary in build quality. The bicycles are bought and sold primarily based on price, and the quality sometimes suffers as a result. I have seen some shockingly bad-quality electric bicycles. There is very little quality control in some Chinese factories, and a significant percentage of the bicycles are shipped broken. A good retailer has to work hard to check the quality of the bicycles he or she is shipping out to his or her customers. Some retailers don't check their products, and it is the customer who ends up doing the quality control by returning the product.

Electric bicycles are still an emerging market, and at the time of this writing, many of the retailers were not short of business. I have experienced extremely busy retailers just not wanting your business unless you want to buy their most profitable product. After-sales service is sometimes neglected.

If you buy an electric bicycle and find a fault, you should contact the distributor and ask for a replacement or refund, no matter how insignificant the fault may seem. Spotting an obvious but minor fault right out the box may be a sign that the overall build quality is poor and that other faults may soon follow. There are typical problems with the gears, chain, pedals, and sprockets. It is often not the electric part of the bicycle but the bicycle itself that is at fault. Even if the bicycle works, sometimes the build quality may be inferior to a cheaper secondhand alternative. If the build quality of a new bicycle is poor, you will find that the frame is too heavy, the handling is poor, and you can't ride "no hands" because the handlebars may oscillate dangerously from side to side. This is why it's good to try before you buy.

### 2.1.4 Resources for Buying an Electric Bicycle

For reviews on electric bicycles in the United States, visit the endless-sphere.com forum. This is a mainly U.S.-based forum, but there are contributors from all over the world. It's an amazing forum, and you can ask anything about electric bicycles and will receive an answer. There is an extensive knowledge base, and I have over a thousand posts on this forum. There are many contributors building their own bicycles there (www.endless-sphere.com/forums/).

For reviews on electric bicycles in the United Kingdom, visit the pedelecs.co.uk forum. This is a U.K.-specific forum, so it will focus on U.K. retailers and dealers. You can ask anything about electric bicycles there. This forum is mainly for commercial electric bicycle owners, but there are a few self-builds (www.pedelecs.co.uk/forum/).

*A-to-B Magazine* features folding bicycles and electric bicycles. It has an electric bicycle buyers guide (www.atob.org.uk/Electric_Buyers%27_Guide.html).

The following is a list of European and U.K. electric bicycle retailers that you might want to look at:

- OnBicycle, www.tebsuk.webs.com/
- Powabyke, www.powabyke.com/
- Urban Mover, www.urbanmover.com/
- Wisper, www.wisper.kellsoft.net/
- Emotive Control Systems, www.emotivecontrolsystems.co.uk/
- Cytronex, www.cytronex.com/
- 50 Cycles, www.50cycles.com/

If there is an annual electric bicycle race in your area, you could look at the bicycles people are using and base your decision on how well they do in the race. The Tour de Presteigne is the U.K.'s premier electric bicycle race (www.tourdepresteigne.co.uk/).

CHAPTER **3**

# Build Your Own
# Electric Bicycle

## 3.1 Why Build Your Own Electric Bicycle?

There are several good reasons why you might want to build your own electric bicycle. Here is a short summary of the main reasons people build their own electric bicycles.

### 3.1.1 Quality and Unique Design

By building your own electric bicycle, you can decide what components to use. This means that you can tailor the bicycle for your needs and desires. You can avoid the cheap Chinese bicycles and electrify a really good bicycle. Often, a secondhand bicycle will have a much better frame than the cheaper electric bicycles. If you build your own electric bicycle, you could choose a chopper-style bicycle, a racer, or a BMX, something you won't find from the normal retailers.

### 3.1.2 Increased Performance

Commercially available bicycles have to follow strict consumer safety legislation, so few retailers sell high-power electric bicycles. In Europe, most electric bicycles have around 250 W of peak power and a top speed of 12 to 15 mph unassisted by pedaling. With these bicycles, hill climbing is impossible, and headwinds will slow you to a crawl. This is changing, and now more powerful bicycles are coming onto the market. If you want an electric bicycle that is faster than this, you have to build your own. If you want a practical bicycle that is a good substitute for a car, then you will have to build your own. Another route may be to upgrade or modify a commercial electric bicycle for more power.

### 3.1.3 Reduced Cost

The profit margin on electric bicycles can be significant, and some bicycles sold for £1,000/$1,500 can be built for £300/$450, so if you are a do-it-yourselfer, there is room to save money. If you build your own electric bicycle, you will research the components and will know that you are getting good value compared with a commercially available electric bicycle. Building an electric bicycle is easy and fun. It is also a really neat project that will teach you all sorts of things about electronics, batteries, and motors. The skills you learn may even get you a job or send you in a new direction. With the skills that you develop, you will be able to perform maintenance and fix the bicycle by yourself. This will save you time and money.

## 3.2  What Do You Want from Your Electric Bicycle?

It is important to have clear and realistic goals before you shell out your hard-earned cash on an expensive kit. To understand your needs/requirements, you should ask yourself the questions listed in Table 3.1.

### 3.2.1  Having Realistic Goals and Expectations

Some newcomers to the electric bicycle arena are filled with unrealistic expectations of what can be achieved; they seem to want a motorcycle, not a bicycle. Owing to the construction of a bicycle frame, there is a limit on the amount of weight that you can put on the bicycle without adversely affecting handling. This naturally will limit the size and weight of the motors and batteries that you can use. The limit on extra weight you can carry on a bicycle is probably around 50 kg/100 lb. However, you probably will find that you are limited by the space available to locate the batteries before you exceed the weight limit. The maximum range you can expect from an electric bicycle, fully laden with batteries, is about 100 miles (with current battery technology). For more range than this, you would have to use a tricycle or trailer to safely hold all the batteries. However, this will increase the cost of constructing your electric bicycle significantly and may make it less flexible in certain situations.

Realistically, though, how long do you want to spend in the saddle? You will get sore after about 1 hour of hard riding. Most electric bicycles are designed for a 20-mile range. The maximum safe speed on a bicycle frame is open for debate. My top speed on an electric bicycle has reached 50 mph. However, bicycle frames are not really built for speeds above 30 mph because they don't have proper suspension. Without enhanced suspension, you will be bounced off the saddle when you hit potholes. Most electric bicycles are designed to do 20 to 25 mph. If you have smooth roads to ride on,

**TABLE 3.1**  Decision-Making Guide for Building the Electric Bicycle You Want

| Question | Answer | Requirements |
|---|---|---|
| How far is it to work (more than 5 miles)? | Yes | Lots of batteries and a frame capable of holding them |
| Do I want to go more than 25 mph? | Yes | Powerful motor and high-discharge-rate batteries |
| Can I recharge at work? | No | Lots of batteries and a frame capable of holding them |
| Do I want to pedal? | No | Motor and batteries rated for continuous use over 500 W power |
| Do I need to take the bicycle on trains often? | Yes | Folding bicycle/frame |
| Does it have to be lightweight? | Yes | High-discharge-rate (expensive) lithium batteries |
| Do I need to carry lots of stuff? | Yes | Basket and rear rack plus batteries cannot take up all the space |
| Do I want to do power wheelies? | Yes | Rear-wheel drive, low gear ratio, small-diameter wheel, powerful high-torque motor controller and batteries, batteries positioned far back on the rear rack |
| Do I want to race or climb steep hills? | Yes | Low gear ratio, small-diameter wheel, powerful high-torque motor controller and batteries |
| Do I want to do burnouts? | Yes | Low gear ratio, small-diameter wheel, powerful high-torque motor controller and batteries, batteries positioned at the opposite end of the bicycle from the motor |

then higher speeds may be acceptable, but traveling at more than 25 mph on a bicycle is very dangerous. Falling off at this speed could be deadly! Wear motorcycle leather gloves, a full-face helmet, and a heavy, thick jacket and trousers if you attempt to beat any land speed records. In fact, you always should wear a helmet and motorcycle gloves on any electric bicycle.

## 3.3  Types of Electric Bicycles You Can Build

Here are some suggestions for types of electric bicycles that you can build, along with photographs showing examples of what each bicycle might look like. For each electric bicycle, there are some build tips and component options you might want to choose.

### 3.3.1 The Long-Range Commuter Electric Bicycle

The long-range commuter electric bicycle is suitable when distances of more than 20 miles are traveled, regularly, that is, such as commuting or if it's not possible to recharge at work. For this bicycle, you need a large frame with lots of space for batteries; for example, a 24-in frame with 26-in wheels is suitable. Figure 3.1 shows a common long-range commuter electric bicycle. This geometry will provide a large main triangle and space for a rear rack to mount panniers. You may need to fill all available space with batteries to get the range required. High-volumetric energy-density batteries are the best choice for this build. Choose pouch cells or primatic cells because they stack better than cylindrical cells. A 50-mile range is not difficult to achieve using this template.

In fact, a larger range could be reached with additional modifications. Mounting of the battery packs will need special attention. Front suspension forks are useful for comfort, but rear suspension might hinder the positioning of batteries. Reliability is key for a long-range commuter electric bicycle, so buy special puncture-resistant tires and slime inner tubes. Make sure that you follow the guides for good electrical wiring and connections in Sections 3.4.16 to 3.4.19. The motor and controller size are less important if only long range is desired and not powerful acceleration. Gear ratios should be selected for efficiency in the 20 to 25 mph zone. A large front chain ring should

**Figure 3.1**    The do-it-yourself (DIY) electric commuter bicycle. This bicycle has a small brushed front hub motor and medium-capacity battery pack, along with a rear carrier, mud guards, powerful lights, and a car horn.

be selected to allow pedaling at this speed. At least one disc brake is essential for stopping power when you have this much battery weight. Luggage carrying can be a problem with all the space taken up by batteries. A rucksack and a front basket are useful if the rest of the bicycle is full of batteries. For a commuter bicycle, powerful lights, mud guards, and a horn are essential. Waterproofing of the wiring and yourself will need careful attention.

If you want extreme range, say, for touring, then to further increase the battery-carrying capability of your frame or bicycle, use either a larger frame such as a tandem frame or a custom-welded frame, add an extracycle conversion or a trailer, or move to a tricycle. You could even store a big battery pack in a rucksack, although for long distances this may not be practical. With one of these modifications, you could build a bicycle with 2 kWh of energy that would take you to 100-mile range. Alternatively, consider using a folding electric bicycle that you can take with you part of the way on a train or bus.

### 3.3.2 The Folding Commuter Electric Bicycle

If you use the train as part of your commute, then a folding commuter electric bicycle would meet your requirements. Figures 3.2 and 3.3 show this type. For crowded commuter trains, you will need a folding frame with either 20- or 16-in wheel size. The smaller the wheels, the smaller is the size the bicycle will fold down to. You can get 26-in folding bicycles, but there is nowhere to put the batteries when the bicycle is folded up. Note that smaller wheels will reduce the ride stability of the bicycle in traffic. In addition, the wheel base (the distance between the wheels) is sometimes narrower on folding bicycles. This affects stability compared with bicycles with larger wheel bases, but the smaller wheel base allows for a more compact folded bicycle. The most popular choice for folding bicycles is a 20-in wheel size because it seems to offer the optimal balance between folded size and rideability. The step down to 16-in wheels doesn't seem to reduce the folded size much because of the need to maintain a stable wheel base (length).

If you are tall, then you can make the seating position bearable by extending the seat to maximum and lowering the handlebars to the minimum. You can make the bicycle more stable by swapping the handlebars over from a normal bicycle. The folding-type handlebars do not allow you to put enough of your weight over the front wheels. Since the space available is limited, a small number of high-power-density batteries are best for this build. The weight of the batteries is less important than the volume, but both should be kept to a minimum. The only space for batteries will be on the rear rack, but most bicycles will not fold up properly if you put a pannier battery there. If you want the bicycle to fold properly, you will need a rear-rack battery holder (see Section 3.4.14).

**FIGURE 3.2**    DIY folding electric bicycle. This one has a large rear hub motor capable of power wheelies, 20-in wheels, and high-discharge-rate batteries. This bicycle also includes front and rear suspension, dual disc brakes, powerful lights, power meter, and speedometer.

**FIGURE 3.3**    Folded up and ready for the train. Sometimes the conversion to electric can affect the folded size. Here, the large after-market forks increase the size of the folded package.

Alternatively, you can just accept that the bicycle will not fold up completely. The fold in the middle is the one that saves the most space because it cuts the length in half. You may find that folding the handlebars or seat is not necessary. Consider upgrading from a retail electric bicycle; the dedicated battery compartment in such a bicycle may be useful. Hub motors are best because they take up the least amount of space because they are inside the wheel, which would have been unused space otherwise.

The folding bicycle is not a stable platform for high-speed performance, so you should select components for a maximum speed of 20 mph. High speeds are best left for larger frames, which are more stable. With small wheels, the gearing is too low for faster electric bicycle speeds. You will need to replace the front chain ring with one from a racing bicycle to be able to pedal at speed. The range is less important because you only need to ride to and from the train station. A high-torque motor might allow you to do power wheelies in a frame such as this with the batteries all on the rear rack. Disc brakes and suspension maybe less important if the bicycle is slower and lighter in weight. V-brakes may be sufficient.

Never bring dangerous stuff onto a train—that is, no gas, no unsafe batteries, and no compressed gases for fuel cells. Keep the wiring neat, or someone might confuse your DIY electric bicycle for a bomb. Watch out for wires that cross the folding portions of the frame, and make sure that wires are not trapped or pulled by the folding movements. Remember to switch off the bicycle before you fold it! Don't obstruct the aisle of the train with your bicycle, and allow other passengers to get off first before you unfold your bicycle. The folding mechanisms gradually loosen up, and this creates instability while riding, so you should check the folding mechanism regularly. Also check the suspension, headset, and battery mountings regularly for loose bolts.

### 3.3.3 High-Power Electric Bicycle

If you just want a really cool electric bicycle that you will enjoy riding or playing with, then this bicycle is for you. You can choose any style of frame, including BMX, folding, mountain bicycle, racer, beach cruiser, or even a kiddie bicycle with stabilizers! See Figures 3.4 and 3.5 for ideas.

Bicycles with small wheel sizes are great fun for doing burnouts or power wheelies. Rear-wheel drive is best for safety, performance, and fun. If you want to do burnouts, put all the battery weight and your body weight away from the powered wheel. This will mean leaning forward on a rear-wheel-drive electric bicycle. It is acceptable to use slick tires and muddy grass to get the burnout started. If you have skill, you can do "donuts," where you use the front brakes and rear motor in combination to tear up mud and carve a donut shape in the ground. If you want to do wheelies,

**FIGURE 3.4**    High-power racing electric bicycle. This model has a medium-capacity 8C-rated battery pack, a powerful overclocked motor, an air-cooled controller, slick tires, and areodynamic dropped handlebars.

**FIGURE 3.5**    A high-power electric bicycle for stunts and power wheelies. This model has a small number of lightweight Lipo battery packs and dual chain-drive motors for high power.

put the weight over the rear wheel and lean back while applying full power. It's possible to build two-wheel-drive electric bicycles for really challenging off-road riding.

For a high-power build, you will need to buy high-power-density batteries. Cost can be kept down if you sacrifice range. High-rate Lipo or LiFePO$_4$ batteries are the best choices. Since the size of the battery pack will be small, you can fit it on any bicycle frame you desire, and positioning won't be a problem. Frames with a large main triangle are best because they provide a low center of gravity and central weight distribution. It may be possible to hide the battery pack in cycle bags for an added stealth effect.

A high-power motor is important, and a high-power controller is even more important. Look into overclocking to reduce the cost of the kit. At least one disc brake is essential for breaking at high speeds—probably two is best. Get a frame with front and rear disc brake mounts and a motor with disc brake mounts too. Torque arms on the motor mounts are important for safety and frame longevity. You can select the motor gear ratio to meet your requirements, low gearing for acceleration and high gearing for high top speed.

### 3.3.4 High Power and Long Range

You might want this bicycle if you really want to push the limits of what's possible with an electric bicycle for racing. Figure 3.6 shows this design. It might be a good idea to pass a motorcycle test or register this electric bicycle as a motor vehicle if your country requires it. The next option is to go all out and build an electric motorcycle that has both power and range.

**FIGURE 3.6**  High power and range. This design has dual disc brakes, suspension, an air-cooled controller, and a large 2C-rated battery pack. Lights and horn are hidden in the basket. Always wear a helmet when riding.

As you increase the range, the weight of the battery pack will increase. A large and heavy battery pack will require better suspension and cause complications with mounting and positioning of the batteries. Braking will become even more critical, and disc brakes are essential front and back. With a large-capacity battery pack, the cost per ampere-hour comes down for having high power because you can get away with buying lower-discharge-rated 2C to 4C cells, for example, $LiFePO_4$, Lipo, or NiMh. The discharge rating of a battery shows you how fast you can use the stored energy and is measured in multiples of capacity (e.g., 3C). This is explained later in Section 3.4.4.

A large rear rack, main triangle, and panniers will be important for carrying batteries. Heat dissipation in the controller and motor will be important during sustained hard riding, and care must be taken when overclocking to avoid damaging components. Have the controller open to the moving air, and consider adding extra heat sinks for added cooling if overheating is a problem.

### 3.3.5 Motored Bicycles

Incidentally, you can put a gas-powered motor on a bicycle. These are called *motored bicycles* and are different from electric bicycles. The best motored bicycle forum can be found at www.motoredbicycles.com/ if you're interested. I built a motored bicycle (Figure 3.7), but it didn't feel as good as an electric bicycle. My 50-cc motorized bicycle could only reach top speeds of 25 mph going downhill, which was way less than my electric bicycle. The motorized bicycle had a two-stroke engine, and the noise it made was tremendous compared with the actual power the motor provided. As a result, I really didn't get the whole effortless electric vehicle (EV) grin feeling that I do with my electric bicycle. I felt like I was riding a lawn mower. In addition, the whole idea of pedaling a gas engine seemed weird. I imagine I got several strange looks. I like my bicycles to be surprising in a good way and not to disappoint.

There is also a whole different set of laws governing the use of gas engines on bicycles, and you could fall foul of these laws, landing you in some serious deep water depending on where you live. I was tailed by police for a block when I was riding my motored bicycle. They probably didn't know what to make of it, and so luckily I didn't get pulled over. I only rode it about three times in total before I sold it. A silent electric motor can be hidden, but a noisy gas engine can't. The motorized bicycles usually aren't as green as they could be either. Gas consumption is usually 100 to 200 mpg, which is actually closer to cars than to electric bicycles. This is about 300 Wh/mile or 15 times more energy than an electric bicycle. On the positive side, a motored bicycle weighs less than a battery-powered electric bicycle and has a

**FIGURE 3.7**    A motored bicycle kit with a two-stroke 50-cc gas engine. It felt like riding a lawn mower.

much longer range. Some people have built nice-looking hybrid bicycles using a gas-powered engine to drive the back wheel and an electric motor for the front wheel.

Most of the motored bicycle kits are not free wheeled; that is, the motor is engaged all the time. You start the motor by pedaling the bicycle up to speed and then dumping the clutch and twisting the throttle at the same time. Once the motor warms up, you can turn off the choke, and then you can just sit tight and motor away. There is only one gear, which is really limiting because engines are only good at a specific speed. With more gears, it probably could reach faster speeds. When you reach a traffic light, you can pull the clutch in and keep the engine ticking over with the throttle. You don't have to bump start the motor again; it will almost pull away from the stop unaided.

## 3.4  How to Build Your Own Electric Bicycle from a Hub Motor Kit

Building an electric bicycle is a lot like building a computer: You buy the components you want, and then you fit them together. The components are in kit form and usually fit together fairly easily. Hub motors are the easiest and most common method of motor attachment, so I will focus on hub motors. Building an electric bicycle requires a few more skills than building a computer, though, because soldering is often needed.

The hardest part about building an electric bicycle is making sure that all the components are compatible with each other. When buying a kit for the first time, it is best to buy the controller and motor from the same vendor because these are the components that are the most complicated to match. Phone the vendor and explain that this is your first electric bicycle kit, and make sure that the parts are compatible. This is an opportunity to check to see if the vendor is a good company with which to do business.

Check if the vendor has any deals or advice on batteries, etc. Keep in mind the type of bicycle on which you are going to be using the kit, and what you are going to use your electric bicycle for. Remember also to buy a hand throttle; hand throttles are cheap and may get thrown in as part of the deal. Most hub motors come already built inside a wheel. This makes it easier for you. Just make sure that you order the correct size wheel for your bicycle. Batteries are not too difficult to match with controllers. You just have to buy the right voltage battery for your controller and a battery with high enough peak current to meet the controller's peak demand.

The order of connection follows this guideline: Everything leads to the controller (see Figures 3.9 and 3.10 later in this chapter). If you bought the controller and motor together, then they should fit to each other easily without having to change any connections. The connections are color coded to make it even easier. The same applies to the hand throttle. If it came with the controller, then connection will be easy. The battery may need the plugs changed or new ones installed if you bought it from a different vendor. You will need to install connectors on the battery and controller that mate. Be careful when installing the battery plugs to keep positive and negative in the right order. If you reverse positive and negative and plug them into your controller, the controller will blow up. To make sure that you get the polarity right, the manufacturer usually color-codes the connectors red for positive and black for negative. The colors will be on the wire's insulation or the plugs. If there is any confusion over wiring, contact the manufacturer or see Section 5.1.1 for polarity detection using a multimeter.

When you buy batteries, buy the battery charger from the same vendor, and make sure that the batteries and charger are compatible. You also may have to wire up the battery charger and charger plug for the batteries (see Section 5.1.1); if there is any confusion, contact the vendor. When you plug the charger into the battery, positive goes to positive and negative goes to negative.

The following instructions guide you step by step in building your own electric bicycle:

- Choose the bicycle you want to convert.
- Decide if you want front- or rear-wheel drive, and buy the kit.
- Take off the old wheel.
- Prepare the new wheel for installation. Put a tire and inner tube on and inflate. If the hub motor was not already built into the wheel, then read Section 3.4.15, and lace it into the rim with spokes.
- Install the new electric hub wheel. Sometimes the forks or dropouts on a bicycle are too narrow to fit a hub motor, especially if it has gear cassettes or disc brakes attached. Most rear dropout sizes are 125 mm/5 in, and most front forks are 100 mm/4 in wide, but some rear hub motors are 135 mm/5.4 in and don't fit in standard bicycles. The solution is to use brute force and bend them. A frame can be bent easily by around 10 mm/0.4 in. A car jack works well for bending bicycle frames (Figure 3.8). Do it slowly in gradual steps, and measure the distance gained between each step. Steel frames are better for this purpose because aluminum can snap. Carbon fiber frames are vulnerable to breaking even without a motor, so never use them. Frames with suspension can be bent slightly too, but this will have repercussions for the suspension because the pistons will no longer be parallel. Some suspension forks cause a fitting problem for hub motors if they have the axle holder in the middle of the fork. This will force you to use spacers to prevent the hub from touching the forks. The spacers will take up too much axle thread and prevent the axle nuts getting a good grip. This will leave the axle unsafe and the wheel too loose.

**FIGURE 3.8**  Using a car jack to increase the distance between the forks to fit a hub motor.

All metal frames can be bent to accommodate a hub motor. You may need to use washers as axle spacers to position the wheel correctly inside the dropouts so that the brake calipers are in the right place for the wheel.

- Install torque arms if you are worried about the dropouts breaking. The dropouts might be vulnerable if you are using regenerative braking, if you are using a front hub motor, or if you are using an aluminum frame (see Section 6.2.2).

- Secure the axle nuts. Use a 6-in-long spanner, and turn the nut as hard as you can by hand. The threads on the nut will strip before those on the axle, so don't worry about damaging the axle. Hub motors usually use fine-metric-thread bolts that can be quite difficult to find, so make sure that you have these sent with the kit. Fine-metric-thread bolts can be found in some specialist hardware stores (e.g., Tool Master). The axles on hub motors are very solid. If you find a nut that almost fits, it can be forced on, and the thread on the axle will cut a new thread into the nut, although it's much better just to have the right bolts to begin with.

- Decide how your batteries are going to be stored on the bicycle. See Sections 3.4.12 to 3.4.14 for help. Install the batteries on the frame, and make sure that they are secure and padded to protect against damage. Lock them to the frame to protect against theft.

- Decide where to place your controller on the bicycle (see Section 3.4.10 for help).

- Make all the wiring connections, as shown in Figures 3.9 and 3.10 depending on which type of motor you have. For tips on wiring connections, see Sections 3.4.16 and 6.2.4 for tidy wiring; for tips on soldering, see Section 3.4.7. The controller connections should all be labeled. You may not need all the connections. There may be two connections for brake cut-off switches consisting of three wires, a three-wire voltage display cable, a two-wire brake light, and a two-wire charger connection cable. These are for scooter style bikes and can all be ignored.

- Cable tie all the wiring to the frame. Make sure that the wires are splash-proof to some extent. Water will not kill them, but it is best to be safe. The throttle cable is the most important one to protect from water. Water in the throttle cable can cause a loss of speed control, sending your motor into warp speed. This is why it's important to have an emergency cutoff switch, or at least a power wire that you can pull out to stop the bicycle. Be careful using tape for waterproofing the throttle wires; it can collect water and have the opposite effect to what you intend.

- Wire up the battery plug and battery charger port (see Section 5.1.1 for help).

- Check everything over before giving it a test ride. Make sure that the brakes are set right and can stop you quickly.

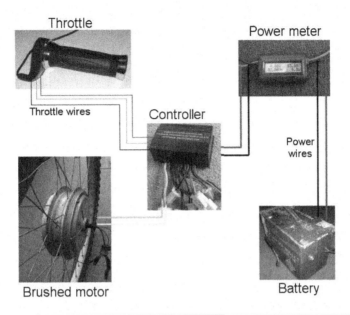

**FIGURE 3.9**    Brushed motor connection scheme. Some of these brushed controllers don't have an on/off switch; instead, they have wires that would plug into the ignition switch on a commercially available bicycle. Just take off the connector, and install an on/off switch there. The controller may not come with instructions so you have to figure this out for yourself.

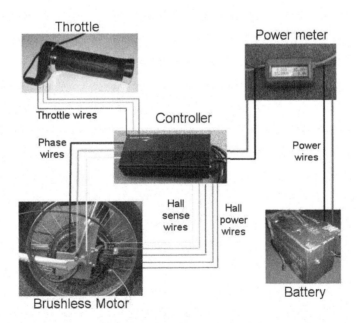

**FIGURE 3.10**    Brushless motor connection scheme. The connectors on the controller are usually all labeled.

### 3.4.1  Build Your Own Electric Bicycle: Kit Resources

For advice on building electric bicycles, visit the endless-sphere.com forum. There is a wide knowledge base there, and you should easily get an answer to all your questions. There are many contributors building their own electric bicycles, including a gallery of DIY electric bicycles (www.endless-sphere .com/forums/).

*Electric Bicycle Kit Suppliers*
- http://ebicycles.ca/
- Solar BBQ, Island Earth (Australian): www.ebicycle.biz/
- Team Hybrid: www.teamhybrid.co.uk/
- AmpedBicycles: http://ampedbicycles.com/
- www.electricrider.com
- www.Itselectric.com
- www.poweridestore.com
- www.emtb.com.au/
- www.electricscooterparts.com
- www.E-Bicyclekit.com
- www.Cycle9.com
- www.stealthelectricbicycles.com.au

### 3.4.2  Electric Bicycle Builders Gallery

Figures 3.11 through 3.14 show what's possible when you build an electric bicycle with a little skill and imagination.

**FIGURE 3.11**   Doctor Bass's 2-kWh full-suspension mountain bicycle with a 100-mile range.

**FIGURE 3.12**  Josh's, aka j-vtol's, creation with a nice looking ABS plastic battery box, disc brakes, and crystalyte kit.

**FIGURE 3.13**  The e-bomber by Jim Winterle. San Antonio, Texas.

**FIGURE 3.14**    Bonzo's ammo box build with headway cells capable of 40 mph.

### 3.4.3  What Determines Power and Speed?

Some people just buy the biggest motor and expect a high top speed, but there is a way to calculate what speed you will get. With more power, you will travel faster, but there are several factors that determine the amount of power that you put to the pavement. Motors are made from coils of wire wrapped around a stator and surrounded by magnets. The number of coils and thickness of this wire will determine the speed and power limits of the motor. If the motor coils are too thin and you attempt to pull too many amperes through them, then they will get hot. Eventually, the heat will build up, and the insulation will melt or the magnets will fail. It is for this reason that motors have power ratings. Each motor will have a continuous and peak power rating. The peak power on most electric bicycle motors is quite high, but the heat will build up if the continuous power is exceeded. These specifications should be available from the manufacturer, and it is best not to exceed them.

The top speed of the motor will be determined by the number of coils of wire and the voltage of the battery that's used to turn it. Motors of the same power can come in different models, where the number of coils will determine whether the motor is designed for high torque or speed. Often a speed/volt specification is used to show what motor speed to expect at no

**FIGURE 3.15**   A cheap but reliable 250-W brushed hub motor, good for 1,000 W peak.

load from a certain voltage of battery. Figure 3.15 shows a hub motor that produces 1,000 W of peak power.

The no-load speed is the top speed of the bicycle. The speed it will reach under no load with its wheel off the ground spinning freely. When you apply a load to a motor and make it do some work, it will use more power to maintain its speed. This means that more amperes are used to provide torque in the motor. In a motor, more amperes will give you more torque, and more volts will give you more speed. Both the amperes and the volts applied to the motor determine the power. The load is created by your weight, the weight of the bicycle, wind resistance, hills, and trying to accelerate. Using an online calculator for the aerodynamics of bicycles, we can calculate the power needed to fight wind resistance (Figure 3.16). This can be used as a guide for selecting the right motor for you to reach a desired speed. As you start to demand more power from your batteries, their voltage will drop slightly; this is called *voltage sag*. When you draw gently from good batteries, the voltage sag will not be noticeable. However, if your batteries don't have a high enough discharge rate for the load, then the voltage sag will slow you down.

In the simplest motors—brushed motors—the amount of amperes that the motor draws will be proportional to the difference between the voltage applied to the motor and the motor's back electromotive force (EMF). The back EMF is generated from the speed of the motor and the speed/volt specification of the motor. In short, if you are going slowly and suddenly turn the

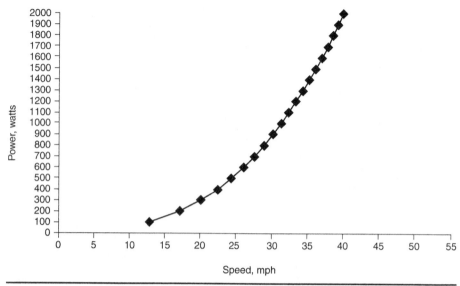

**FIGURE 3.16** Constant-speed versus power graph (estimated) for a typical mountain bicycle–style electric bicycle on flat ground with no wind. You can use this graph to estimate the power requirement to reach a desired speed. The graph assumes constant speed, so in reality you would need a long road to reach this top speed. It may be better to have more power to allow for acceleration up to the desired speed.

throttle wide open, you will be drawing a lot of amperes, but if you slowly build up speed, the voltage difference and ampere use will be less. This is how a brushed motor controller controls the speed of the motor—by altering the voltage supplied. For a detailed description of how brushless motors work, see Section 4.1.1. Thus your speed could be determined by the motor, the batteries, or the controller. For how to determine which factor is most limiting, see Section 6.1.4 on overclocking. You have to make sure that the motor, batteries, and controller that you buy are all rated for the power that you want and all rated for the same power. It's no good buying expensive heavy components for their capabilities if you're not going to use them fully. See Chapter 4 for more information on components.

### 3.4.4  What Determines Capacity and Range?

The capacity and discharge rate of your battery pack will determine how far you can go and how fast you can go safely without damaging your battery (Figure 3.17). Therefore, you need to carefully and conservatively consider two things when deciding the size of your pack. Power consumption on an electric bicycle varies depending on how fast you go and how much you pedal. A good rule of thumb would be 20 Wh/mile and 1 extra Wh/mile for every extra mile per hour of speed you do.

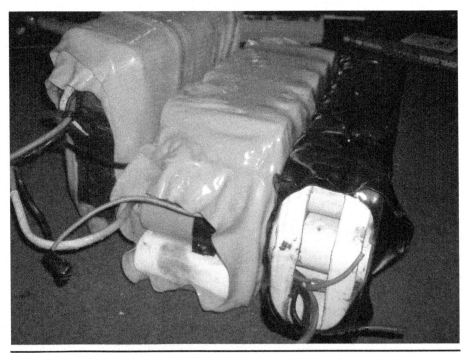

**FIGURE 3.17** Lots of nickel–metal hydride (NiMh) cells. Rating at 7 Ah and 2C peak discharge means that these batteries are good for only 14 amperes.

- Normal cyclist baseline speed is around 14 mph.
- 15 Wh/mile for slow speeds (15 to 20 mph) and normal pedaling.
- 20 to 25 Wh/mile for average speeds (20 to 25 mph) and relaxed pedaling.
- 30 Wh/mile for fast speeds (>30 mph) and no pedaling.
- At high acceleration and 30 mph, pedaling makes very little difference to the speed.

### Battery Sizing Example: Pack Capacity

Let's assume that your commute is 5 miles, you want to be able to go 30 mph, and you don't want to have to charge at work. From Figure 3.18, your energy consumption will be 35 Wh/mile. Therefore, your commute would need a battery pack size of at least

35 Wh/mile × 10 miles = 350 Wh

### Battery Sizing Example: Pack Discharge Rate

You must size your pack so that it's big enough to cope with both peak and continuous power demands and not just the range you need. The discharge rating of a battery shows you how fast you can use the stored energy and is

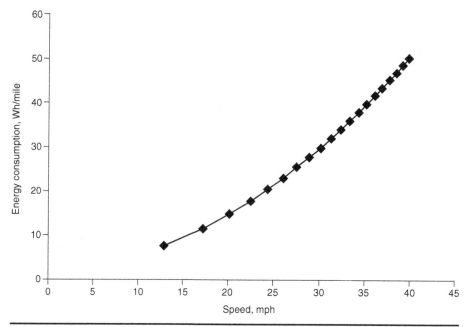

**FIGURE 3.18** Battery depletion (estimated) at different speeds for a typical mountain bicycle–style electric bicycle on flat ground with no tail- or headwinds.

measured in multiples of capacity (e.g, 3C). This is the multiple of the capacity of the pack, in amperes, that you can safely draw on to make power. The units for *C rate* don't make sense from a scientific point of view, but the notation has stuck around because it's an easy measurement to use. Using Figure 3.16, you can see that you need 900 W to travel at 30 mph. Therefore, to travel at 30 mph, you either need a 350-Wh capacity battery pack rated for 3C continuous discharge rate or a 900-Wh capacity battery pack rated for 1C continuous discharge. The C rating can apply equally to amperes and ampere-hours as it does with watts and watthours. A 2C discharge of a 10 Ah battery pack will be 20 amperes. Batteries usually will have a continuous and peak discharge rating, and if you stay inside these limits, then the batteries will last the number of cycles that are claimed by the manufacturer, and you will be very happy. If you exceed these limits, however, you will receive fewer battery cycles from your big battery investment. The biggest mistake people make when building an electric bicycle is not buying a powerful enough battery pack for the given discharge rate and then wrecking it because they "ragged it" without knowing. This is why it's so important to get a power meter. Also, if you notice that your bicycle lights go dim when you accelerate, then you are pulling too much from your batteries. Depth of discharge and speed of recharge also will affect battery life.

**Other Things to Consider When Sizing a Battery Pack**

There also will be a recommended C rate for charging a battery. Therefore, another reason for buying a large battery pack is so you can charge at a faster rate. Maybe you want to go out in the evening or go shopping right after your long commute back from work. With a larger battery pack, you can put energy into your battery at a faster rate. Batteries with low C rates like to be charged slowly, and this is relative to their size. Having a larger battery pack generally will make your cells lasts longer.

Next, you need to take into account such things as cold winter temperatures and pack aging, which will reduce the capacity of your batteries. To be conservative, you should at least add an extra half capacity to cope with such situations. If you buy a higher-capacity battery pack, then you don't need such high-discharge-rated cells, so it may be best value for the money to buy a 900-Wh 1C pack for the job rather than a 350-Wh 3C pack. The disadvantage will be that the 900-Wh pack will weigh more.

The main use for fast-discharge-rated packs is to save weight. Then you can safely drive fast and for much longer, up to 26 miles at 30 mph. This kind of electric bicycle is a seriously awesome kit that is much better than any cheap Chinese eBay bicycle, and the only way you can currently get one is to build it yourself. However, batteries are expensive. A good, lightweight 900-Wh battery pack currently costs about £400 to £500/$600 to $750. Battery pack purchase is a personal decision for the owner, and it pays to know all the facts. Think of the big battery pack as an investment. Each time you cycle to town for a booze-up, you are saving a round-trip taxi fare. You just need to do that 20 times, and your investment has paid off! (Disclaimer: This author does not condone cycling under the influence of alcohol.)

### 3.4.5 Weight Distribution in Electric Bicycles

With heavy batteries and motors, it is not always possible to distribute the weight evenly. Often, the best that you can do is to balance the weight between the front and back of the bicycle. If the bicycle's weight is out of balance, then some weird riding characteristics can result. It's important to be aware of these and avoid them. For example, putting all the weight out back on an electric bicycle can result in dangerous front-wheel skids when braking in the wet and going downhill. Front-wheel skidding is very scary. Watch out for this problem if you have a rear hub motor and lots of batteries in rear pannier bags. Use your rear brake to stop you, and hopefully it is a disc brake.

If you do find yourself in a front-wheel skid, you have to release the front brake and then reapply it once the skid has subsided in much the same way that an automatic braking system (ABS) in a car works. Keep the bicycle going straight while you are doing this. Front hub motors also can create front-wheel skids but only through wheel spinning in wet conditions going uphill. Wheel spinning is not really a problem and is easily controllable. Too much forward weight distribution can cause a different problem—sore hands. With all the weight up front, the tires will be weighed down and provide less bounce. This magnifies road surface bumps and gives the rider sore hands even on bicycles with front suspension.

The ideal weight distribution in electric bicycles is low down and evenly placed across the bicycle. It is for this reason that the Tidal Force electric bicycle has its battery in a special front hub–mounted case, which makes the bicycle look like it has two hub motors. This keeps the weight low and balances it out between the front and the back. The disadvantage of hub-mounted motors and batteries is that they are not supported by the suspension—they are unsprung weight. Unsprung weight increases wear on the bicycle and can result in such problems as broken spokes or rim damage. Other electric bicycle manufacturers try to spread the load out by putting the cells inside the frame, but this has the negative side effect of difficult maintenance access. Keeping the weight as low as possible is important for handling, especially when jumping curbs or doing wheelies. A top-heavy bicycle will feel more unstable and will require more effort when turning. A novice rider will weave side to side when riding a top-heavy bicycle. Again, be careful who you let test drive your bicycle—some people will not expect a bicycle to weigh so much and will drop it.

### 3.4.6 Brakes on Electric Bicycles

Brakes on fast electric bicycles get used a lot more than those on slow-moving regular bicycles, and therefore, it is important to make sure that the brakes are strong and reliable. They could save your life! A bicycle with good brakes also will be more fun to ride because you will feel more in control, and you will feel safe in applying full power of the electric motor (and skidding is fun).

Disc brakes are the best types of brakes because they are strong and maintenance-free. An electric bicycle simply must have at least one disc brake and preferably two. Hydraulic disc brakes are slightly better than cable disc brakes on the rear where you have long and winding cable runs. Long bendy brake cables will have lots of sponginess to them, which is cured by changing to hydraulics. Rim brakes can be strong when adjusted properly, but they require lots of readjustments because of the force involved in stopping a fast bicycle. Readjustments may be needed as often as once a week. This can be a real pain if the brakes are hard to reach, under a pannier bag full of batteries. Rim brake pads also will get used up really fast on a powerful electric bicycle (as often as once every few months). In a wet environment, rim brakes will be next to useless, and if your wheel is slightly buckled, then rim brakes will rub and cause drag. With rim brakes, your wheel has to be exactly centered with no sideways wobble, no vertical wobble, no egged rim, and proper dishing. If any of these defects exist, rim braking will be weak, and you may be able to brake effectively only on a small fraction of the total circumference of the wheel, causing inconsistent braking force. With disc brakes, this stuff doesn't matter, and this is much better for people who have to lace their own hub motor because it's very difficult to get it exactly right.

Unfortunately, on some big rear hub motors, such as the X5, there isn't enough space on the axle to fit a disc brake rotor without some difficult modifications. The same is true for chain-drive bicycles with rear-wheel sprockets; these usually take the space where the rear disc rotor would sit. Some hub motors have a disc brake option. Definitely choose one of these, if available, even if you don't use the disc brake capability right away because you may find yourself wanting the option later. With electric bicycles, another type of brake is available—the electromagnet brake. This is discussed in Section 6.2.1.

On any type of bicycle, usually the front brake is the most important brake because all the weight of the bicycle is transferred to the front during braking. In fact, if the braking force is strong enough, the back wheel may lift off the ground. For fast braking, therefore, it is advisable to have a front disc brake and to put the hub motor on the back. With a rear hub motor and batteries in the back, all the weight is on the back, which keeps the back wheel on the ground and allows for more braking options on the rear wheel. With some big rear hub motors, a disc brake may not fit, so braking is limited to a rim brake, which is useless in wet conditions. The disadvantage of this setup is that in wet conditions, the maximum forward stopping power can cause front-wheel skids when going downhill. Because the rear rim brake is insufficient, the rider uses the front brake to stop, and the lack of weight over the front wheel causes it to skid. This is why it is important to have disc brakes on both the front and the back when using a fast and heavy electric bicycle. It also shows the importance of slowing down when it's wet on the roads! The opposite setup is to have a front hub motor and a rear disc brake. With this setup, braking in the wet during downhill travel is safer, but stopping times in general are increased.

### 3.4.7 Step-by-Step Guide to Soldering

Soldering is the key skill to master when building an electric bicycle. There is no substitute. Soldering is not difficult, but it does at times seem to require four hands with which to hold the two wires to be soldered, the soldering iron, and the solder! If you are not good at soldering now, by the end of building your first electric bicycle, you will be an expert. If you have never soldered before, don't worry—it's easy. Go out and buy a soldering kit, and have a go. It's possible to learn the art of soldering in a few hours, and once you have learned it, you will never forget—it's just like riding a bicycle.

*Soldering Made Easy*
- Start with two pieces of thin wire, a hot soldering iron, and some solder.
- Wire-strip the tips of both wires to be soldered.
- Touch the iron to solder to melt a blob of solder onto the iron. The iron is now said to be *tinned*.
- Touch the wire to be soldered to the tinned soldering iron to heat it up. A thin wire should heat up in seconds. The liquid solder on the iron is very good at heat transfer.
- Tin the wire by touching it to solder while it is receiving heat from the iron. The solder should flow quickly to all hot parts of the wire (three things must touch).
- Tin the other wire to be joined by repeating the two steps immediately above.
- Touch the two wires together, and heat them with the iron to melt the two tinned coatings into each other (three things must touch).
- Hold the wires together for a few seconds, and wait for the solder to set.

**Tips for Soldering Electric Bicycle Stuff**
Electric bicycles use big, high-current wires that are difficult to solder because they dissipate heat so quickly. To solder these big wires. you will need a big (high-power) soldering iron. I recommend a 50-W iron for big wires. The best soldering irons have a power-selection dial so that you can turn them down so as not to overheat delicate stuff. Avoid soldering gun–style irons that are not designed for continuous use; they break too easily.

If you are stuck with a 35-W iron, in a pinch, you can still solder high-current wires if you follow this method:

Wire-strip the tips of both wires, and then pinch a few of the copper strands to the side.

Because you have reduced the number of wire strands, the heat dissipation will be less, and you should be able to solder the selected wires together.

You then will have the two wires connected by a few strands but with lots of loose strands around the outside. You then repeat the process with the other strands until all the strands are connected in a single connection. Or, if you're lazy, you can just tape the other wires together and hope that the connection is good enough with all the wires touching.

### Tips for Soldering Multiple Wires

If your soldering iron came with a holder, you can put the iron into the holder backwards, and this will free up one of your hands for holding other things, such as solder and tricky wires. Alternatively, you can rest the iron on the edge of a table and with a weight on top of the handle to hold it in place while you solder.

### Tips for Joining Multiple Wires

Soldering more than two wires together is difficult in the extreme because as you solder the third wire, the first wire connection can melt and fall off. To help with multiple-wire connections, you can use *terminal blocks*. Solder the wires together in pairs so that their ends are pointing in the same direction. Then use the terminal block to join these two soldered wires together with other wires. In this way, you can connect four wires with one terminal block, six wires with two terminal blocks, eight wires with three terminal blocks, etc. This technique is really good for running the bicycle lights off the main battery because it allows you to connect a switch, front light, rear light, battery power indicator, and other accessories very easily and neatly (Figure 3.19). See Section 6.2.4.

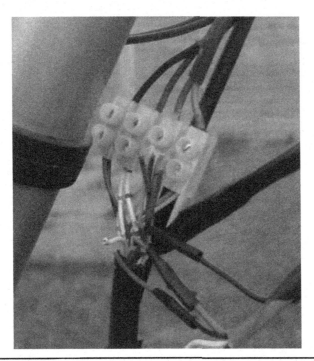

FIGURE 3.19   Using terminal blocks to connect many wires is easier than soldering them. Terminal blocks can be used for connecting sets of lights, switches, and a horn.

### 3.4.8 Mounting and Securing Things to the Electric Bicycle

Mounting and securing things to your electric bicycle is important. This was the activity that I found myself spending the most time doing. The electrical parts of the bicycle are the most reliable; it's the mechanical parts that experience wear and are likely to fail first. What follows are descriptions of some of the methods I have found helpful when building electric bicycles. Table 3.2 lists the basic tools you will need to accomplish these tasks.

**TABLE 3.2**    Tool Requirements for Electric Bicycle Building

| Tool | Description |
| --- | --- |
| Wire strippers | Pistol-style is easier. |
| Soldering iron | I recommend a temperature-selectable 50-W iron. |
| Solder | Leaded solder is best for soldering thick power wires. |
| Pliers | For cutting and griping. |
| Plastic electrical tape | |
| Multimeter | This is essential. |
| Spanner set | |
| Screwdriver set | |
| Allen key set | |
| Glue gun or epoxy | |
| Duct tape | |
| Cable ties | |

### Padlocks and Chains

Padlocks are good for three reasons: They are secure from theft; they are solid and don't break under vibrations caused by riding; and they are easy to lock and unlock. A padlock and a chain are good for holding batteries on your electric bicycle (Figure 3.20). However, they can be a bit bulky sometimes, and too many of them make it look to thieves like your electric bicycle is valuable. Use black duct tape over the locks and chain for more stealth.

### Drilling and Tapping

Drilling and tapping are very useful for making brackets for disc brakes or battery holders. For adding rear disc brakes to a non–disc brake frame, I have used 10-mm-thick aluminum tool plate, drilled and tapped to hold the brake caliper (see Section 6.2.2). For securing batteries, I have drilled and tapped angle iron onto military surplus ammunition boxes to hold them (see Section 3.4.14 and Figure 3.20). The drill and tap method is used when there is either no room to fit a nut on the other side, when it has to be flush with

**FIGURE 3.20**    Battery pack held on with padlocks and tapped holes.

the surface, or when the presence of a nut would make the union less secure. This method is used commercially to secure lockable boxes to rear racks so that a thief can't just undo the bolts from the other side. On disc brakes, it is used because there is no room for a nut because the disc is flush with the surface of the bracket.

### Cable Ties and Reusable Cable Ties

Cable ties seem to be what's holding most DIY electric bicycles together these days! Cable ties come in lots of sizes, colors, and quality! They are mainly good for tying back cables, though, and will break if used to secure batteries. You can use them to hold your controller, but it's best to have something a little more secure against thieves. Sharp edges on things will snap them. Cable ties are made of nylon, and they degrade slowly over time, becoming brittle. Reusable cable ties are best used for securing luggage-like D-locks onto the bicycle frame. The release mechanism makes them less flexible for use and less secure. There are metal cable ties, but these aren't as strong as they sound, and the locking mechanism is prone to slipping.

### Duct Tape

This stuff is great! I have it wrapped all over my electric bicycle. Thick, heavy tape is great for holding lightweight, soft things together or for covering surfaces or waterproofing. It is also good for hiding unsightly wires.

### Hose Clips

The next step up from metal cable ties is hose clips. These can be opened and wrapped round things to make strong connections. These are good for DIY torque wrenches using 10-mm spanners.

### Coil Springs

Springs are good for joining things that need a bit of elasticity. Use steel springs that have the ends looped round. The loop is used to secure the spring to an object or to another spring using nuts and bolts. If you wrap the spring around something that needs to be tied and then bolt the loops together, it makes a very strong tie that also has some elasticity. This type of setup can be used to hold batteries to racks.

### Welding

If you know someone with an arc welder, then the sky's the limit in terms of battery mounting and frame design. An arc welder is a great tool to have if you have a steel bicycle frame and steel ammunition boxes holding your batteries. To prevent melting holes in stuff by accident it is helpful to clamp a third piece of metal underneath the two pieces you are joining. This way when you melt the two metals together the piece underneath will collect the melt and stop it from escaping and leaving a hole.

### 3.4.9  Choosing a Bicycle Frame

With electric motors you have plenty of power, so there is no need to worry about making the lightest possible bicycle. Bicycle weight doesn't really affect speed at all. Weight may affect acceleration a bit, but the weight of the rider usually far exceeds the weight of the bicycle, so it's not worth worrying about. Instead, you should concentrate on getting a frame that is solid, reliable, and built like rock. It's going to take a lot of punishment. A steel frame is better than an aluminum frame when you use a powerful hub motor because the entire driving force of the motor acts on the axle groove in the dropouts. This does not occur in a regular bicycle, and therefore, regular bicycle frames are not really designed to cope with such force. In some aluminum frames, the axle of the hub motor can grind away at the dropouts; some hub motors even have been known to completely eat away the dropouts and spin inside them. When this happens, it is bad because all the wires in the motor get wrapped around the axle and ripped out of their mounting points. This can be difficult to repair.

Ideally, there should be no obvious visible signs that your bicycle is an electric bicycle that is hiding £500/$750 worth of batteries and £400/$600 worth of fast motor and controller! This is a tall order, and the best starting point is a large bicycle frame with lots of space to put stuff. A big triangle in the frame is very useful, so avoid frames with the lower "female" crossbar. The suspension in dual-suspension mountain bicycles will get in the way, so it's best to avoid these too. If you choose a frame with 20-in wheels, then you may find that the gearing of the pedals is set up for speeds of only up to 15 mph. If you go any faster than that, you will find yourself unable to contribute by pedaling, which will detract from your enjoyment of the ride. To help with this, you can swap chain rings over from a racer bicycle if you have the right tool. This will allow you to pedal up to 20 mph with the biggest 60T cog on the front and a 12T cog on the back. The small wheels really limit pedaling at high speeds. With 26-in wheels and the highest bicycle gearing, you can pedal up to speeds of 30 mph. Pedaling becomes a bit of a moot point at these speeds anyway because the amount of energy you can contribute is a tiny fraction of the amount of energy that your motor puts out to maintain the speed.

### Practical Tip

Find out where all the abandoned bicycles in your city go to get recycled. This place is likely to give you great bargains and have a wide variety of components that you may want.

### 3.4.10  Positioning Your Components on the Frame

Building an electric bicycle is not too difficult, but probably the hardest thing, after choosing your components, is positioning them on the bicycle. There are several practical considerations to address when designing your electric bicycle, and often you will find that decisions will be a result of compromise between these factors. The main practical challenges to an electric bicycle are rain, heat, weight, theft, aesthetics, reliability, and convenience. When you look at purpose-built electric bicycle frames, there are always dedicated spaces designed to store the battery and the controller. On a regular bicycle frame, you don't have these purpose-built sections, so you have to make them. Even off-the-shelf electric bicycles will have insufficient space if an upgrade is desired. Battery storage is one of the hardest tasks the electric bicycle builder will face. The batteries have to be firmly secured and protected from theft and bumps, but they also may have to dissipate heat effectively and be low enough to avoid making the bicycle top-heavy. The controller will have an even bigger heat-dissipation problem and may need cooling airflow, but at the same time you need to protect it from rain and theft. If you have several battery packs onboard, then you will have a whole heap of big, thick wiring and connectors that needs to be accessible but also needs to be covered from rain and for aesthetic reasons.

There are several places to store the controller; it is easier to place than the batteries because of its light weight and size. Possible locations are inside the main triangle, behind the main triangle, on the rear rack, or in a pannier bag. Most controllers are waterproof to some extent or can be made waterproof with some silicone sealant, duct tape, or a bag. Therefore, rain is not the biggest concern—theft is the main threat. Some controllers cost up to £300/$450, but still, many people just cable-tie their controllers to the frame. I have been told by police that most bicycle thieves are junkies with a pair of pliers as their only tool. They simply crouch down by a bicycle and cut through the cable locks one wire at a time. To an observer, they may look like they were unlocking their bicycle. Even if passersby do spot a bicycle thief, they usually will not want to get involved. The best theft-prevention strategy seems to be to never leave your bicycle unattended for long periods.

As a builder of electric bicycles, you can't completely design out theft. What you can do is to increase the number of tools and time that a bicycle thief would need to steal your bicycle or components. You can also dissuade a thief from trying to steal your bicycle by hiding its true value or fitting an alarm. Since a pair of pliers is the most common bicycle theft tool, it would be good to stay away from fixing methods that can be defeated by pliers, such as cable ties. The best method to secure a controller is to padlock it to the frame or rack (Figure 3.21). This requires a thief to carry a larger, more obvious theft tool such as bolt cutters or a saw, and this is unlikely.

**FIGURE 3.21**    Controller padlocked to the frame with good airflow for cooling and duct tape for splash-proofing. Plastic tape is used to seal the seams in the case.

Another way to avoid theft is to hide the components. Put the controller in a zipped-up bag. The zippers often can be padlocked together to prevent opening. You need to be able to switch the controller off when the bicycle is parked. Poke a hole in the bag to allow a finger to go through to reach the off switch so that you can switch the controller off when it is inside the bag. Monitor the temperature of your controller after the initial ride to be sure that heat buildup is not an issue with the bag. If you're going to do any racing, remember to move the controller to a more ventilated location. It's good to pad your controller so that it receives less of a knocking when riding. Cable-tie some rubber matting between it and the surface to which it is locked down.

Stealth involves concealing the fact that your bicycle is electric. You may want to do this for a variety of reasons: to look like an athletic cyclist, to conceal your bicycle's true value from thieves, or to avoid getting stopped by police if the laws in your state/country prohibit electric bicycles. It's possible to make a bicycle so stealthy that even a top cycling dude could not tell if it was electric.

*Stealth Tips*
- Cover wires and shiny things that look out of the ordinary with duct tape.
- Spray paint shiny hub motors black or coat them in oil and grease so that they don't stand out.
- Hide wiring in a bag.
- Paint such things as the battery box and controller the same color as the bicycle so that they look like they are original factory fittings.
- Choose motors with no noise, that is, only direct-drive hub motors, no chain drive, and no geared motors.
- Small motors are less conspicuous than large ones, so overclock a smaller motor rather than buying a high-power motor to get the power you desire.
- Choose a rear hub motor, and hide it behind a set of pannier bags.
- Hide things inside regular-looking cycle equipment if possible.
- Use a common frame type with which people are familiar—for example, no recumbents, tandems, tricycles, or strange-looking folding bicycles.
- Consider hiding hub motors inside a set of wheel disc covers.
- Keep bicycle lights low power like regular bicycle lights; 20-W beams raise suspicion.
- Don't oil your chain; let it rust. People will hear your pedaling effort, which will be much louder than your motor.
- Be very courteous and friendly to pedestrians you pass along your cycle route. Say "Good morning" to them, and they will look at you and not your bicycle.
- Start pedaling and slow down when you pass people; use the motor again when you are clear.

### 3.4.11 Battery Storage and Positioning

The best place to store batteries is down low in the panniers or in the frame's main triangle. A rear-rack carrier is essential for carrying batteries, so make sure that the bicycle either comes with one or is able to accept one later. Steel rear racks with strong support legs are better than the seatpost paddle kind because they can hold more weight. Seatpost paddle racks have to be reinforced to carry batteries. I once gave my 14-stone/200-pound friend a lift on my electric bicycle's rear carry rack. It was a steel rack built into the bicycle. When this rack rusted to bits, I swapped it for an aluminum rack rated for 25 kg/55 lb, and it failed inside a week under only the weight of my 15-kg/33-lb batteries. Avoid aluminum racks; they are no good for carrying batteries. The ratings on racks seem to be way too optimistic, and lots of people have had problems with broken aluminum racks.

Pannier bags are good for storing luggage or batteries. With three-bag design panniers, you can use the space in the top bag to store your wiring and controller and the bags on either side for your batteries. This is a good design because it gives you a low center of gravity, which really makes a difference when turning or hopping curbs. Be sure to buy strong panniers because they will receive lots of wear from rubbing between batteries and frame; extra padding may be needed. Few panniers are strong enough to support 15 kg/33 lb of batteries without breaking, and frequently you will have to provide additional support. To further secure the batteries from theft, you can attach them to the frame's rear dropouts.

Try to avoid putting heavy batteries on top of the rear rack because it will raise your center of gravity, and this will make handling difficult. It also will damage the bicycle rack if there is no padding between them. You will quickly get used to your bicycle weighing 40 kg/88 lb, but be careful who you let test drive your electric bicycle—most people will not expect a bicycle to weigh so much and will drop it or fall over. Make sure to warn people about the hand throttle when you hand the bicycle to them because some people will twist the throttle unintentionally, and the bicycle will do a backflip.

### 3.4.12 Making a Main Triangle Battery Bag

A main triangle battery holder is one of the simplest DIY solutions, but it is really suitable only for hard-cased batteries because there is no strongbox to protect vulnerable pouch cells. Inside the main triangle is a good place to store batteries; it's central and has a middle center of gravity.

There are several ways to secure batteries in the triangle. The best would be to weld a custom frame, but this is not an option for most people. Instead, you can strap and tie the batteries to the frame. A good method is to hang the batteries from the crossbar in a shoulder bag. The bottom of the bag can be tied to the lower frame to stop it from waving around. The bag will need to be secured using something stronger than just flimsy material, though. The best choice is to use plastic-coated wire metal fence material available from your local garden center (Figure 3.22). Don't use bare metal chicken wire because it risks shorts in the batteries. The bag will offer a bit of insulation, but be aware that the movements of the bicycle over bumps will cause the heavy batteries to rub against whatever material you have used. To avoid a possible catastrophic short circuit against the metal frame of your bicycle (Figure 3.23), you will need to use a strong padding against abrasion. You can use

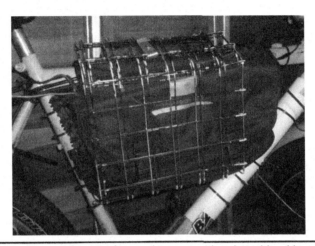

**FIGURE 3.22**    Plastic-coated metal fence wire support bracket with rubber padding.

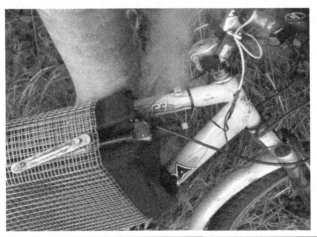

**FIGURE 3.23**    Not like this! Without rubber padding, battery insulation is eroded by grinding against the metal frame on bumpy roads. This caused a short circuit across the frame. Sparks melted the steel frame, and heat destroyed some of the cells. Battery performance was significantly reduced. A fuse does not prevent this kind of incident! This shows the importance of good battery padding, housing, and insulation.

thick rubber matting material to absorb mechanical shock from the bumps, and you also can use hard plastic PVC pipe material to insulate the frame.

*Making the DIY Main Triangle Battery Holder*
- Gather the parts you need; these include rubber matting, shoulder bag, plastic-coated metal fence material, PVC pipe, padlocks and cable ties, plugs for charger connection, Anderson connectors for power, fuse holders and fuses, and battery cutoff switch.
- Gather the tools you need (see Table 3.2).
- Wire up the battery using Anderson or other main power connectors. Be sure to put a fuse holder with a fuse on the positive power wire as close to the battery as possible. Add a plug for charging if one is not already connected by the battery manufacturer. Add the battery cutoff switch and place it somewhere accessible in case you need to cut the power in an emergency. If your batteries are not already wired up with charger port and power connection and you are unsure how to wire them up, then consult the retailer or see the guide in Section 5.4.13 and the multimeter polarity check tip on Section 5.1.1.
- Pad the shoulder bag with thick rubber matting material, and then put the batteries in the shoulder bag.
- Zip up the shoulder bag so that the wires are the only thing coming out.
- Put the bag's satchel flap over the crossbar, and fasten it so as to hang the bag from the crossbar.
- Bend the plastic-coated metal fence material in half along its longest side to form a rectangle. For aesthetics, you might want to spray paint the wire a different color to match your bicycle or the bag.
- Bend the rectangle of fence material around the bag and crossbar to form the shape of your support bracket.
- Wrap the frame with the PVC pipe where the batteries might touch it. For aesthetics, you might want to spray paint the PVC pipe to match the frame.
- Move the support bracket up to the bag and crossbar. Stuff the support bracket with rubber matting under where the batteries might touch the frame. This gives more padding to protect against abrasion.
- Tension the fence material support bracket against the crossbar, and use small padlocks or hose clips to fix it together.
- Cable-tie the bottom of the support bracket to the lower half of the main triangle at each end of the bag to stop it from swinging.

### 3.4.13 Making a Theft-Proof Battery Holder

To further secure the batteries from theft, you can attach them to the frame. To do this, you have to store your batteries in a metal case that you can pad-

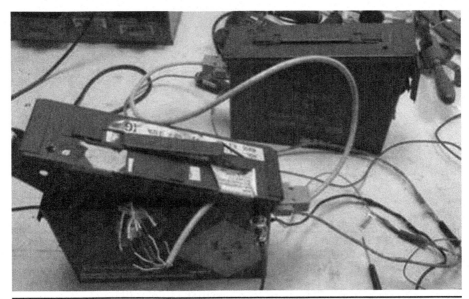

**FIGURE 3.24**   Battery boxes made from strong ammunition boxes. These battery holders were designed to fit in a double pannier set and sit on either side of the rear wheel. Remember to peel the warning stickers off the boxes and coat them in paint or duct tape; otherwise, people might get the wrong idea. You don't want to find that your bicycle has been the subject of a controlled explosion by the bomb squad!

lock to the frame (Figure 3.24). The same padlock(s) can be used to lock the case and lock it to the frame. You also could use a chain to support it from the frame off the crossbar or rear rack. If you have an arc welder, then you can use that instead and skip some of the steps.

*Making a Theft-Proof Battery Holder*
- Gather the parts you need: these include a metal box, padlocks, a chain, padding, plastic-coated metal fence material (optional), wiring strain-relief grommets, panel-mount plugs for charger connection, Anderson connectors for power, fuse holders and fuses, and a battery cutoff switch. Suitable metal cases can be found in your local supermarket, hardware store, or army surplus store—just go shopping and keep your eyes open. Metal cookie tins, steel breadboxes (Figure 3.25), toolboxes, and ammunition boxes all have potential. Thick steel is best for protection. A well-padded ammunition box is the best choice (see Figure 3.24).
- Gather the tools you need (see Table 3.2). In addition, you also will need an electric drill and drill bits and a file.
- You can get padlocks and chain from a DIY or hardware store. The thicker the better, but 4-mm chain works well. Cut the chain to the right length.

**FIGURE 3.25**    Battery box and battery showing rubber matting used for padding. The blue thing on top of the battery is the BMS. This is a breadbox battery holder.

Hopefully, it will be hard to cut so that a thief will have a hard time. Sometimes you can just bend the chin link back and forth with pliers until it breaks. Use the minimum amount of chain so that it's tight; then any thief will have a hard time trying to get access and leverage.

- Drill holes in the metal box so that it can be padlocked shut. Obviously, don't drill the case while the battery is inside! The holes should be just slightly bigger than the padlock so that it fits easily. Drill the holes, and then file the edges smooth. Make sure that there are no sharp edges to the box that might puncture your battery or cut into the insulation surrounding wires. Clean any metal filings from the case before you insert the battery.
- For a professional job, you can panel mount the charger port or any other thing, such as battery management system extension wires, an on/off switch, etc.
- **Important:** Pad the case with thick rubber to prevent damage to your batteries from moving (Figure 3.26).
- Put the batteries in the metal case. Run the wires through the drilled holes, but use grommets for strain relief and to stop the metal from wearing away at the insulation and causing a short circuit.
- Wire up the battery using Anderson or other main power connectors. Be sure to put a fuse holder with a fuse on the positive power wire inside the metal battery box. Add a plug for charging if one is not already connected by the battery manufacturer. Place the battery cutoff switch somewhere accessible in case you need to cut the power in an emergency. If your batteries are not already wired up with charger port and power connection and you are unsure how to wire up your batteries, then consult

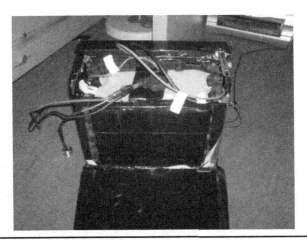

**FIGURE 3.26**    Battery snugly fitting inside a well-padded box.

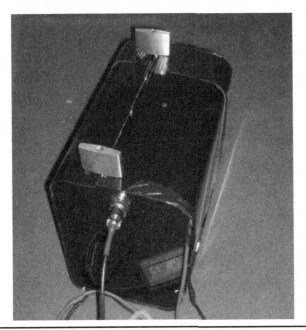

**FIGURE 3.27**    Securely locked battery box. This is then hung from the locks using chain over the crossbar.

the retailer or see the guide to wiring a new battery in Section 5.4.13 and the multimeter polarity check tip in Section 5.1.1.

- Once the battery box is ready, padlock it to the frame of the bicycle. The same padlock(s) can be used to lock the case and lock it to the frame. If the box is strong enough, such as an ammunition box, you can use the chain to hang it from the crossbar or rear rack by the handles. If the box doesn't have a strong handle that will support its weight, you can use wire mesh to support the box from underneath. Or you can use both chain and wire mesh, as shown in Figure 3.28.

• You can put the batteries inside a shoulder bag or a pannier bag to hide them from view and add waterproofing.

**FIGURE 3.28**    Battery box padlocked to the frame with a steel chain. The box is supported at the top by the chain and at the bottom by the wire mesh holder. The shoulder bag is then added to hide the box.

### 3.4.14  Making a Rear-Rack Battery Holder

It's harder to make a heavy battery stay on top of something than hang underneath something. The rear-rack battery holder is useful, although on folding bicycles or dual-suspension bicycles with a single crossbar design and no triangle to put the batteries, there is more space on top of the rear rack for a battery than there is available on the sides. You can't fold a folding bicycle with big pannier batteries on the sides, and aerodynamics is better without large side panniers. Positioning a battery above the wheel is a more symmetrical weight distribution for a single battery than hanging it from one side of the rack. The key to mounting a heavy battery on the rear rack is to add suspension and dampening between the battery box and the rack so that the mountings don't shake themselves to pieces with the vibrations of riding (Figure 3.29). The rack has to be strong, and the bolts need to be checked regularly to make sure they are not vibrating themselves undone. Arc welding would be very useful for this project, but it is possible to do it without.

Seatpost-style paddle racks tend to break under the weight of a battery. They must be reinforced from underneath with diagonal bracing. You have to anchor the battery to the rack at all four corners to stop it from moving around. It is possible to buy off-the-shelf lockable rear cycle carriers for the

**FIGURE 3.29**    Battery box mounted to a rear rack with theft-resistant iron brackets that are secured with padlocks and a chain to the frame and steel cables to the rack. Suspension is provided by an inflated inner tube and rubber mat between the box and rack.

rear rack (Figure 3.30), but they are made of plastic and won't last very long with batteries inside them; also, they are easily broken into.

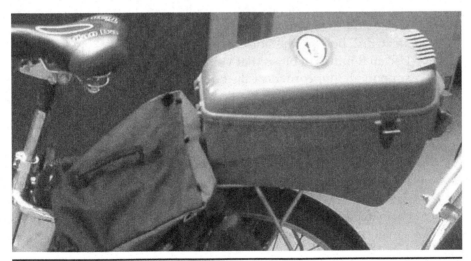

**FIGURE 3.30**    A commercially available plastic rear cycle carrier is inadequate for the task of carrying batteries. Plastic hinges can be broken easily by thieves, and the lock can be defeated easily. The battery inside would need to be well padded for protection or it might get damaged by shocks from the road. If the cells were vulnerable pouch cells, then they would need to be inside a second, stronger box for protection.

*Making a Rear-Rack Battery Holder*

- Select a solidly built metal box for your battery. An ammunition box is best.
- Source all the other components you will need, including 3-mm-thick angle iron, 6-mm bolts and a tap with ISO M6 thread, 6- and 5-mm drill bits and a drill, a padlock, some chain, an inner tube, a rubber mat, cable ties, steel brake cable, and fine-threaded nuts and bolts.
- Drill holes for mounting the recharge socket and feed-throughs for power wires. File the holes smooth, and use strain-relief grommets to prevent short circuit of the battery power wires on the sharp corners of the holes.
- Obtain two angle-iron brackets, and drill 5-mm holes in them for mounting onto the box. You must have at least two holes on each to make them secure. The holes have to line up with 6-mm holes you will drill in the bottom of the metal box.
- Position the angle iron along the sides of the box so that the brackets run parallel to the sides of the rear rack but slightly wider. The rack needs to sit between the angle-iron brackets without the rack hitting them. Use the drilled angle iron as a template for where to drill the 6-mm holes in the box to make them line up. Pilot drill the holes first with a 5-mm drill bit.
- Tap an ISO M6 thread into the 5-mm holes you have drilled in the angle iron.
- Use the 6-mm bolts to connect the angle iron to the bottom of the metal box. The bolts can only be tightened or loosened with access to the inside of the box where the heads of the bolts are located. This is crucial for security. A thief cannot unscrew the bolts from the underside because there are no nuts to loosen.
- Next, drill holes in the sides of the angle iron for locking to the rear rack. Use an 8-mm drill bit so that you can use 6-mm padlocks. Drill holes to support all four corners of the box.
- Next, cable-tie the rubber matting and inner tube to the middle of the rear rack so that it stays in place when the bicycle is bouncing around on rough terrain.
- Duct tape the underside of the battery box to smooth over any sharp edges that could puncture the inner tube.
- Place the battery on top so that it is supported by the inner tube, and secure it from the sides using the padlocks and springs or hose clips. Use the chain to padlock the battery to the frame as well as to the rear rack so that a thief cannot simply steal the whole rear-rack assembly.

The commercial rear cycle carrier uses threaded brackets in a similar way, but instead they run under the rear rack to hold it in place. This semi-permanently fixes the carrier to the rack. The metal rear-rack battery holder described here is a superior design for electric bicycles because it provides extra suspension for the heavy battery and allows the battery to be removed easily, if necessary, for recharging.

### 3.4.15 Tips on Wheel-Building Hub Motors

Sometimes you will need to build a hub motor into a new rim either to replace a broken rim or to change wheel size. You probably could write a whole book on wheel building, but I will offer just a few tips to get you started. These instructions may be the wrong way to do it. I didn't read any books on how to do this. I just gave it a try and it worked!

My daily commuter bicycle has a custom-laced 12-kg/26-lb motor with a heavy battery box on top of it. It has worked nicely and has broken only a few spokes, so it's really not that difficult to maintain. The hardest part is finding the right-sized spokes for a given rim and hub motor combination. There are several online calculators that can help you with this. In addition, 14G spokes are available on eBay made to any length. It's not really possible to cut and thread spokes yourself because it requires a special machine, not just a simple thread-cutting die. Once you have the spokes, the next difficult part is getting started. You can't put spokes in some directions when the wheel is already under tension, but it's also difficult to do if the wheel is flopping around owing to lack of tension. Some people use a wheel-building rig for this reason to support the wheel and rim while you insert the spokes.

- Wheels can be built from lots of different spoke combinations, so it's not always necessary to buy new spokes if you have lots of different-sized spokes already lying around. A large and small spoke set can be used in combination to replace a spoke size somewhere between the two.
- If your spokes are a little short or a little long, then you can make up a few millimeters/inches by changing whether the spokes are threaded from outside in or from inside out.
- The spoke size needed will change depending on how the wheel is spoked. With the double-cross pattern, the spokes need to be longer than with the radial pattern.
- Remember to get the right size spokes and the right number of them. Also remember that 26-in wheels usually have 32 spokes and 20-in wheels usually have 36 spokes, but this can vary.
- When using powerful hub motors, a thicker gauge of spoke should be used or the spokes all will slowly snap. Thus 10G (3.2-mm) spokes are best for this purpose instead of the normal 14G (1.7-mm) spokes.
- Sometimes rear wheels require different-sized spokes for each side of the hub because the wheel needs to be dished to keep the rim central on the axle if there is a big gear cassette on one side.

- You also can dish the wheel by positioning the spoke butts all on one side. This is useful for the rear wheel.
- Don't be afraid to bend spokes to get them into position. They will straighten out later when you tension them.
- With powerful hub motors it's best to get wide steel rims and large, over-sized tires. Narrow aluminum rims won't last.
- The normal spoke pattern used for bicycles is the triple-crossover pattern, where each spoke crosses three other spokes on the same side of the wheel, once at the hub and two more times. This is recommended by wheel builders for strength, but it won't work with big hub motors in small rims because the spokes have to be angled too much at the rim. In this case, for example, with an X5 motor in a 20-in rim, you must use the radial pattern.
- If you want to upgrade the spoke size on a motor, you may have to drill the holes wider.
- It is possible to drill extra holes in a rim to change it from 32 to 36 holes, but it's best to use strong steel rims for this purpose.
- There are different types of nipples available with threads in different places. If the spokes are a few millimeters/inches too short with one set of nipples, then they might be a match with another set. Electric hub wheels seem to have different nipples than regular wheels (Figure 3.31). They are longer with more thread, making wheel building easier.

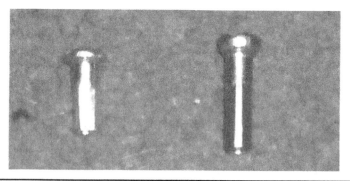

**Figure 3.31**    Electric bicycle wheel nipples are longer and make wheel building easier. (Left) Regular bicycle wheel nipples. (Right) Electric bicycle wheel nipples.

*How to Lace a Wheel*

- Get the spokes ready. Calculate the right size spokes for the hub-rim combination, and then buy the spokes.
- Get the rim ready. Rims can be bought with the number of holes to fit any hub motor. Make sure that the number of holes in the rim matches the number in the motor.

- Get the hub motor ready.
- Get a spoke key (Figure 3.32).

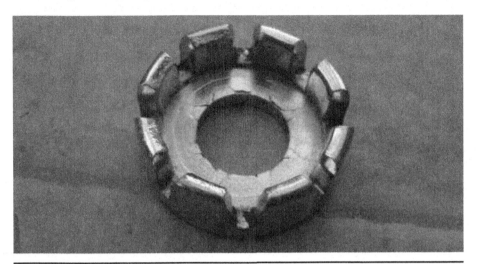

**FIGURE 3.32**   A spoke key is essential. Round ones are best.

- The best way to start lacing a wheel with a cross-pattern is to build up from a simple radial pattern. Thread one spoke every four holes on the same side in a radial pattern. The spokes will be too long and stick out from the rim.
- Turn the hub around inside the rim until the spoke nipples hit the rim holes and the spokes go tight (Figure 3.33). Make sure that the nipples are all tightened to the same extent.

**FIGURE 3.33**   Twist the hub in the rim until the slack is taken up.

- Next, lace a few spokes in the other direction, one every four holes. After a few spokes in the opposite direction are threaded, the spokes will support the hub and hold it in position inside the rim.
- Next, thread the spokes on the other side of the wheel in the same way, one every four holes. The one-in-four method is used because there are two directions in which the spokes go, on each side of the wheel. Thread the spokes through from the already laced side of the wheel so that they dangle free and can be maneuvered into position without clashing with the already laced spokes (Figure 3.34).
- Put the wheel into the bicycle fork.
- Tighten all the spokes a bit more. Remember to dish the wheel slightly if there is a gear cassette or a disc brake on one side.
- Tighten up all the spokes in the correct sequence to make the wheel true. The wheel needs to be centered in the radial direction as well as true with the axle at 90 degrees to the plane of the rim. To do this, spin the wheel and look for sections where the wheel sticks out sideways. Where the wheel angles to one side, tighten the spokes on the other side to pull the

**FIGURE 3.34**  Top side of the wheel is fully laced. To spoke the other side, you have to drop the spokes through the holes from the already laced side. Otherwise, it's not possible to maneuver them into position because of all the other spokes already in the way. Note how the spokes exit on the bottom of both hub flanges. This is to dish the wheel so that the rim is central despite being offset by the disc brake mount on the bottom.

rim over. Also look for sections where the wheel sticks out in the radial direction. Where the wheel sticks out of center in the radial direction, tighten the spokes on the other side to pull the rim to the center. Don't overtighten the spokes. Only go to finger tightness with the spoke key.

- Make sure that the threads on the spokes aren't extending out into where the inner tube will be or they will cause a puncture. Cut the tips off the threads with pliers if necessary.
- Wrap lots of thick rim tape round the rim to protect the inner tube from spoke punctures. Install the inner tube and tire to finish the wheel. You can get special straight inner tubes that don't require the wheel to be removed when fixing punctures. These are the best choice if you find removing the wheel a hard task.
- After the bicycle has been ridden for a few miles and the wheel has been worn in, re-tighten the spokes.

### 3.4.16  How to Do Electrical Connections and Wiring

If you don't do your electric bicycle wiring right, you will find yourself stuck at the side of a road somewhere pushing a 60-kg/130-lb lump and wondering what went wrong. It's important to build reliability into an electric bicycle from the beginning. Electric bicycles can be very reliable, but those built by amateurs will break down. Wiring problems are common faults on an electric bicycle. They are not usually serious or costly, but they are annoying. There are three potential failures to avoid with your wiring, and they are heat buildup, loose connections, and short circuits. The rest of this section is devoted to examining these problems in further detail.

Amperes generate heat. The more amperes you have, the thicker your current-carrying wires need to be. The resistance of a wire goes up as its area goes down. Therefore, everything that is in the path of your current needs to be a specific minimum thickness. The classic mistake is the *paired-wire short circuit*. This is where a pair of wires carrying current to and from a battery are too thin, and the excessive current generates heat when squeezing through the thin-wire *pinch point*. The heat buildup melts the wire's insulation, and the wires make contact, causing a short circuit across the battery. If you have other batteries in series, then the symptom of this failure will be a sudden drop in power, or the power may cut out altogether. To avoid this problem, you must make sure that every current-carrying wire that goes into your bicycle is fit for purpose. The practical tips below can help as a guide. The problem usually will happen only when you are pulling lots of amperes for long periods. Thus, when you are racing to get somewhere important is when it will strike! Heat dissipation also will factor into the problem. If the wires are packed into a bag, then any warming will build up and might cascade into a full-scale melting. This is similar to how a fuse works. However, it is not just wires that need to be thick; it is everything in the current's main path. Therefore, switches, relays, fuses, fuse holders, and connectors all need to be rated to cope with the amperes you demand if they are in the main current path.

I once entered an electric bicycle race with my commuter electric bicycle without any time to upgrade for the race. As a result, I had to stop three times to replace melted components: three fuses, three relays, and a controller overtemperature shutdown (I didn't win). If you are ever in any doubt about the thickness or ampere rating of a wire or component, you can always try the touch test. Just wire it up, ride for a bit, and then see if it gets warm to the touch. If it gets warm, then replace it with something thicker.

You can measure the power dissipation in your wires with a multimeter. Set it to millivolts, and touch the probes at either end of a main power wire with only wire between the probes. Go for a short ride and check. This will tell you the voltage drop across the wire. It should be only a few millivolts. Multiply this voltage by the amperes from your power meter, and the result is the power dissipated as heat in the section of wire between your two probes. Hopefully, the number will be below 100 mV, which at 30 A would produce wiring heating of only 3 W. You have to decide what wiring losses are acceptable to you. As a guide, a 25-W soldering iron gets very hot, so keeping your wiring below this is desirable.

Loose connections are bad because they can cause sparks, power loss, and short circuits. A loose connection results when you haven't done your connections properly and something has come loose. This can be the result of solder aging, rusting, poor soldering technique, vibration, movement, or a bad cable-crimping technique (Figures 3.35 and 3.36). If a main current connection is intermittent, you will find that your electric bicycle will work only if it is tilted in a certain direction, and you will experience lack of power. An intermittent connection will cause sparking at the connection, and this could damage the component. If a plug has come loose or something is blocking contacts in a switch, then the plug might fuse together or the switch might weld shut. A loose connection that's flapping around also might inadvertently make a connection with something else that it shouldn't do. This could cause a short circuit and fry a component or something else unexpectedly. A short circuit results when positive and negative terminals are allowed to touch with no load or resistance in between. When this happens on an electric bicycle battery, only the cell's internal resistance and the resistance of the wire will limit the current. Things will get very hot and start to melt metal, although hopefully the system will have a fuse that melts first.

FIGURE 3.35    Sloppy electric bicycle wiring causes loose connections and short circuits. Here you can see too many wires not tied down and bare metal not insulated.

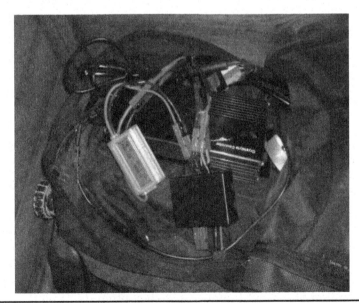

FIGURE 3.36    Here you can see tidy wiring all tied down with a good junction box. The controller is padlocked to the rack, with rubber padding under the controller to protect it from bumps.

**Practical Tips**

Table 3.3 shows what wire sizes are recommended for different current flows. The current shown in the table is the maximum continuous current and is intended as a guide for wiring a house. Current demand on an electric bicycle fluctuates, and there are usually periods of inactivity or low speed where the wires can cool down. Therefore, electric bicycles can use slightly smaller wire sizes if desired.

For 20-A controllers, its best to use at least 2.5-mm-diameter wire, and for 40-A controllers, use at least 3.5-mm-diameter wire.

**TABLE 3.3**  Wire Size Guide for Copper Wires

| American Wire Gauge (AWG) | Wire Diameter, mm | Maximum Continuous Amperes |
| --- | --- | --- |
| 3 | 5.83 | 75 |
| 4 | 5.19 | 60 |
| 5 | 4.62 | 47 |
| 6 | 4.11 | 37 |
| 7 | 3.67 | 30 |
| 8 | 3.26 | 24 |
| 9 | 2.91 | 19 |
| 10 | 2.59 | 15 |
| 11 | 2.3 | 12 |
| 12 | 2.05 | 9.3 |
| 13 | 1.83 | 7.4 |
| 14 | 1.63 | 5.9 |

Source: www.powerstream.com/Wire_Size.htm.

Fuses are very important on your electric bicycle, and they will save your batteries from sudden death if there is a wiring fault that causes a short circuit. Always have a fuse on every battery on the bicycle. Fuses should be positioned inside the battery case so that if the insulation on the power wires were to fail by being caught on the sharp-edged battery box, then the fuse would be in the path of the short-circuit current. Make the edges of the wiring outlet holes smooth and free of burrs. Don't be tempted to put the fuse holders in accessible locations; they should be inside the battery box. If lots of fuses blow, then you have a problem—either a short circuit that you were unaware of or you are pulling too many amperes. Fuses are not a good indication of current flow because they often don't blow until the current exceeds twice their rating. Whether a fuse blows or not is very time-dependent. Blowing fuses is not an accurate way of estimating the effects of current flow on batteries. Use strain-relief grommets on the holes where the power wires exit the battery box. If the box is made of metal, these holes can have sharp edges that will cut the insulation on the wires and cause a short circuit. Hot glue can be used to provide some instant strain-relief grommets if the wire is already in place and you don't want to remove it.

It can be annoying having to half disassemble your electric bicycle to change a main power fuse each time the wires from your lights get wrapped around the handlebars and short circuit. The best thing to prevent unnecessary blowing of main power fuses is to install lower-value fuses on all accessory power lines. Usually it's best to put a 5-A fuse on the accessories line that goes to the handlebars. The best fuse holders to use are waterproof in-line blade fuse holders; these will hold standard automotive fuses up to 30 A. It's a pain when a main power fuse blows because it often melts the fuse holder along with it and requires a rewiring job on the batteries. Melted fuse holders can be confusing to diagnose, especially if you have battery management system (BMS) cutouts to rule out first, so always examine or replace a fuse holder after a main power fuse blows.

To neaten up your handlebars and hide some of those unsightly wires, you can buy a single 20-core cable and feed all your handlebar wiring down that. The 20-core wire can replace the separate wires needed for power to the lights, rear lights, throttle, power meter, and horn. All you need is a terminal block for a breakout box at each end (see Section 6.2.4).

### 3.4.17 Choosing Quality Connectors

Most electrical connections on your bicycle will be soldered. Ensure that your soldering technique is good by reading the guide in Section 3.4.7 and practicing. Where such things as chargers, battery packs, and controllers might need to be plugged and unplugged, you will want to use connectors. There are a variety of different connectors you could use on an electric bicycle. You must make sure that you make the connection properly and that the connector is rated for the amperes and volts that will be used. The voltage rating on things such as connectors and switches relates to the insulation and shows how high the voltage can go before problems such as arcing occur. Usually, the rated voltage of standard insulation will be much higher than the typical voltages used on electric bicycles. Even just electrical tape will work fine for electric bicycles.

The current rating is usually much more important than the voltage rating. The voltage rating on switches usually just means that you will have a reduced lifetime for the switch if you surpass the rating. This might not be important if it's a cheap switch in a nonessential area. Figures 3.37 through 3.40 show examples of plugs and connectors that are useful on electric bicycles. In addition to voltage and current rating, there is reliability and build quality to consider. Many of the connectors used on cheap Chinese electric bicycle controllers and motors are awful and will break with continual use. Sometimes the first thing you do with a new kit is upgrade the connectors to something more reliable. Some wiring connection problems can be very difficult to diagnose because the symptoms might be similar to those of more serious problems. If one pin in a Hall sensor plug breaks, you may spend a long time trying to determine if the problem is in the controller or the motor, and you might cause further damage by attempting to "fix" things that aren't broken. In addition, some connector problems can cause a knock-on effect that can damage other components.

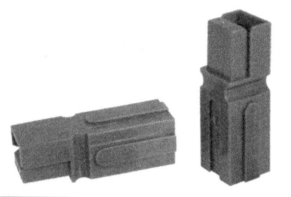

**FIGURE 3.37**    Anderson power poles. My connector of choice.

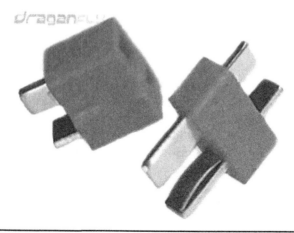

**FIGURE 3.38**    Deans connectors.

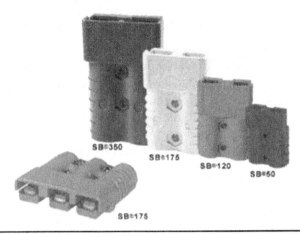

**FIGURE 3.39**  Multipole Anderson connectors.

**FIGURE 3.40**  A Euro connector or "kettle lead."

**Practical Tips**

Anderson power poles (see Figure 3.37) are the electric bicycle builder's connector of choice because they can handle lots of amperes and are compact. Anderson power poles come in lots of sizes, but the 30- and 45-A, 7.9-mm-wide sizes are the ones of interest to us. The connectors are assembled by crimping them or soldering the pin onto a wire and then pushing this through the back of the housing. A very small screwdriver is helpful to poke the pin into the housing. It is good practice to seal the back of the housing with hot glue or epoxy to make the connection last longer and seal any bare wire. The power poles are unisex in that there are no male or female ones; they all can connect with each other.

Anderson housings come in lots of different colors, so remember to always select red for positive and black for negative wires. This way you can avoid an accidental short circuit. The power poles have dovetail slots along their edges, so they can stack together to form multipole plugs. This also will help prevent accidental short circuits because you can create plugs that will fit together in only one orientation. Anderson power poles are available on eBay. Anderson also makes multipole connectors for this purpose (see Figure 3.39).

Deans connectors are popular with remote control (RC) hobbyists. They are multipole connectors with orientation-specific male and female connectors (see Figure 3.38). Other connector types are also available. In a pinch, Euro connectors or "kettle leads" can be used (see Figure 3.40). They are available commercially but are bulky, difficult to mate, and support only limited amperes.

Another good practice is to label all your wires. After you've had a few pints or it's dark, it can be difficult to tell the difference between identical-looking connectors in a spaghetti-like mess of wiring. A label maker can be invaluable for labeling all your connectors.

### 3.4.18  Choosing Quality Switches and Relays

Conducting high currents is a demanding task for a contactor, and switching high voltages can cause powerful sparks. A switch can be used to control an electric bicycle. The switches on an electric bicycle need to be higher quality and much bigger than other switches if they are in the path of the main current. Often there are special requirements for switches, such as switching between multiple poles or multiswitches. These specifications are abbreviated on switches and relays to double pole (DP) and double throw (DT) (Figure 3.41). For example, a double-pole switch can be used to switch a connection between two separate circuits. A double-pole, double-throw (DPDT) switch might be used to switch both positive and negative terminals from a battery between separate charge and discharge circuits.

Relays (Figure 3.42) are big switches that can be activated by remote by a smaller switch. This allows for high-current or high-voltage circuits to be controlled by small, compact controls from a distance. Relays are useful in electric bicycles because you don't want to have thick high-current wires running to your handlebars and back. You always need to have at least one high-power switch on an electric bicycle for an emergency cutoff switch in the event that your controller goes haywire and locks up with the throttle wide open.

Other applications for switches are battery-switching arrays, and relays can be used for special applications such as electric brakes. For electric bicy-

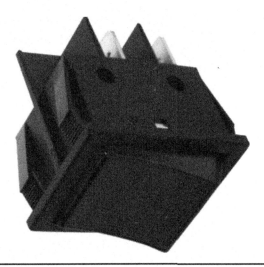

**FIGURE 3.41**    A 16-A double-throw rocker switch. Use both sides together for 30 A.

**FIGURE 3.42**    A 20-A automotive relay.

cle applications, normal switches of above 15 A will do (see Figure 3.41). For racing or high-performance electric bicycles, however, these switches could be melted by the high current. For this purpose, you can gang multiple poles together on DPDT switches. For switching higher voltages, you need special vacuum contactors (Figure 3.43) to stop arcing across the contacts because of oxidization of the air. These are more expensive. For this reason, on high-powered electric bicycles, it's best to minimize the number of switches used. Controllers often have a special electronic switch built in, so there may be no need for any main current switches.

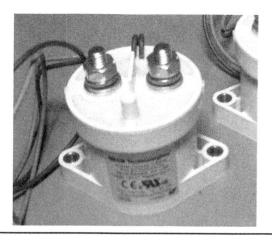

**FIGURE 3.43**    A Kilovac contactor 500-A, 320-V dc vacuum relay.

### Practical Tips

A high-amperage emergency cutout switch doesn't have to be an expensive and bulky contactor. You can just attach a pull cord to the back of one of your main power wire plugs. If there is a problem, pull the cord, and the main power connection is severed. You obviously have to anchor the other side of the connector so that the force pulls the plug out of the socket.

If a switch is required to switch live circuits on/off, then the voltage specification will need to be higher.

### 3.4.19  Making Wires Tidy: Junction Boxes

The typical home-made electric bicycle will have lots of wires carrying current from different battery locations and sending it to the controller, lights, horns, etc. It is essential to do a good electrical job with the wiring because otherwise your electric bicycle will be unreliable and cause much misery. You have to use switches or plugs on your battery network for many reasons. You may want to disconnect batteries if there is a problem, or you may want to switch batteries between series or parallel configurations for performance or recharging. For these reasons, you have to use either a switch box or a junction box. The advantage of the switch box is that it's faster to use and can be made user friendly. The advantage of a junction box is that it's easier and more flexible.

Figure 3.44 shows a series/parallel battery-switching circuit. I used this circuit for charging a high-capacity NiMh battery in series and then switched to parallel for discharge. This was necessary because NiMh cells cannot be charged safely in parallel and because in series the voltage is too high for the controller.

**FIGURE 3.44**  Plan for a series/parallel battery switching circuit. For charging in series and discharging in parallel.

It is important to mount switches inside a box to protect them from movement during riding. Junction boxes (Figures 3.45 and 3.46) or plug boxes are easier because they are basically just a load of plugs connected together. It still pays to get it right, though, to ensure a reliable ride. There are several professional junction boxes that you can use. The circular plastic junction boxes are good for up to 4 +Ve/−Ve plug pairs. For more junctions than this, you need a strip of terminal block. For electric bicycles, use either the 20- or 30-A blocks. The plugs can be left outside the case or can be brought inside for more secure connections.

Once you have your switch or junction box, it is important to tie it down to stop it from moving around and to stop plugs coming loose or switches

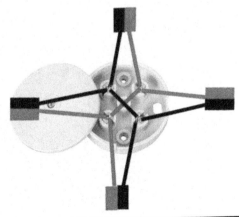

**FIGURE 3.45**  Inside a junction box: electric bicycle wiring using a terminal block in a case.

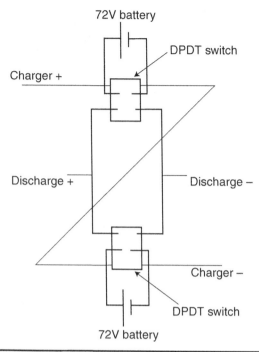

72V battery

DPDT switch

Charger +

Discharge +

Discharge −

Charger −

DPDT switch

72V battery

**FIGURE 3.46**  A junction box designed to connect two batteries in parallel, several accessories, a controller, and a power meter.

being switched. You can put all your unsightly wiring in a dedicated pannier bag or box. Don't put anything loose in the wiring bag/box. It's important to keep leads short so that they don't catch on things and pull out (Figure 3.47).

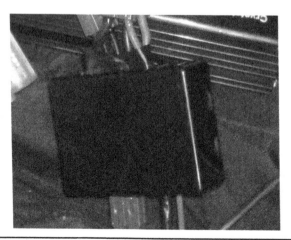

**FIGURE 3.47**  Electric bicycle wiring using a junction box.

# CHAPTER 4

# Electric Bicycle Components

## 4.1 Motors

### 4.1.1 How Motors Work

Motors are made from coils of wire wrapped round a stator (iron core) which is free to rotate and surrounded by magnets (Figure 4.1). Motors are very similar to electromagnets; they use the same attraction and repulsion forces. The trick with a motor is to switch the direction of the forces to keep the motor moving after its initial revolution. There are two main types of motors used in electric bicycles, and these are brushed and brushless. Sometimes brushed motors are called *direct-current (dc) motors* and brushless motors are called *alternating-current (ac) motors*. The difference between these motors lies in how their power is switched to keep them spinning. In a brushed dc motor, the switching is done mechanically by brushes that sweep the sides of the hub and swap the polarity of the power as the motor spins round. In brushless ac motors, the switching is done electronically by the controller, and the motor position is sensed using electromagnetic sensors called *Hall sensors* to obtain the position of the motor. The main difference from a performance perspective is potentially slightly higher efficiency with the brushless motor (5 to 10 percent more efficient). It is for this reason that you will find that most of the electric bicycles on the market have brushless motors so that they can go farther on a small battery. The mechanical switching in brushed motors causes sparks as the brushes make contact. This gradually causes erosion of the brush contacts. Brushless motors can be built for higher power requirements because they don't have the problem of spark erosion.

The power in a brushless motor is controlled by pulse-width modulation (PWM), which is a scheme in which the controller selects how long to

**Figure 4.1**    Inside a Crystalyte X5 brushless motor. The stator and windings are in the middle, fixed to the axle, which is fixed to the bicycle and doesn't move. The wheel with the magnets attached revolves around the outside. The side coverplates of the motor contain sealed bearings that support and hold the two parts in position while allowing them to rotate.

switch the power on during each phase. If full power is selected on the hand throttle, then during each phase the power is switched on all the time. If partial throttle is selected, the power to the phase is switched on for only a short time.

### 4.1.2  What Speed Will It Do?

The speed and power limits of a motor will be determined by several things, including the number of coils, the thickness of wire, and how quickly the heat can escape from the motor. When you apply full power to a motor with its wheel off the ground using a battery and controller, the motor will achieve a top "no-load" speed. The top speed of the motor under real conditions usually will be around 20 percent less than this no-lead speed. No-load speed is proportional to voltage. You can calculate the speed per volt of the motor and scale it up to find the top speed at any voltage. The voltage is limited, though, by the specification of the controller. However, the actual speed in real-world situations will be determined by the power of the system and the load.

### The Balance of Load versus Motor Power Is What Determines Real Speed

Calculation of real-world maximum speed under real-world load is complicated, and you may wish to skip it and just look at the simplified speed versus power graph in Figure 3.16 to size your motor. Motor efficiency increases with speed to the point where at its top speed it will be most efficient. However, the power the motor provides reaches a peak and then drops off as the speed approaches maximum speed. At slow speeds, the motor is very inefficient. This is called the *motor characteristics curve* (Figure 4.2). The load on an electric bicycle is mainly the energy used to fight wind resistance and to fight gravity when going uphill. Because of wind resistance, the force required to obtain a speed increases with the square of the speed. This is called the *load curve*. When you apply a load to a motor, it will use more power to maintain its speed or will lose speed if the power is not increased. You need to know both the load curve and the motor characteristics to calculate the speed that will be reached.

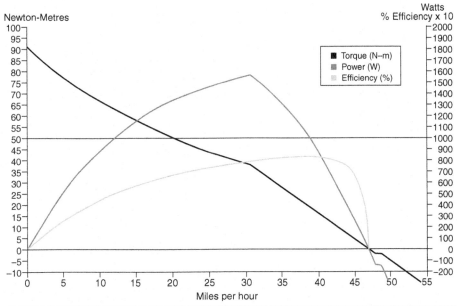

**FIGURE 4.2**    Motor performance characteristics for an X5303 hub motor at 48 V in an electric bicycle with 26-in wheels and a 50-A controller at full throttle. The motor efficiently increases with speed to the point where at its top speed it will be most efficient. However, the power the motor provides reaches a peak and then drops off as the speed approaches maximum. At slow speeds, the motor is very inefficient. (Source: Motor simulator at www.ebikes.ca.)

You can do either a force balance or a power balance to work out the speed. Where the graphs of power versus speed for the load and the motor cross, this is the speed that will be achieved. That is, where Figure 3.16 and Figure 4.2 cross, that is the speed that will be obtained under those conditions when steady state is reached (Figure 4.3). This is much more complicated than we need to make motor/battery purchasing decisions, and as soon as the grade or wind speed changes, the load curve will change, and a different speed will be reached. In practice, it will take a while to reach this steady-state speed. Therefore, motors should be sized slightly higher to allow for the increased power required by acceleration up to the desired speed.

The motor simulator at www.Ebikes.ca and the speed-power calculator at www.kreuzotter.de can do the work for you. Use the speed-power calculator to calculate the power required to reach your desired speed, and then choose the motor with at least that power. Play around with selecting the different motors, voltages, and wheel sizes to see what they do to the position of the efficiency peak and the power peak. It's best to operate the motor at a speed between the efficiency peak and the power peak.

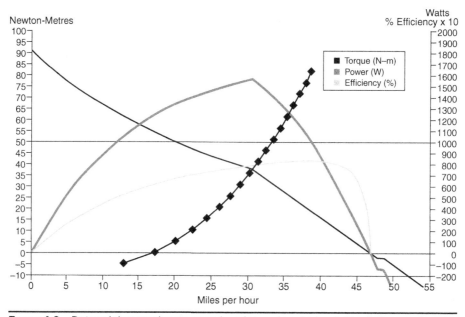

**FIGURE 4.3**  Determining maximum speed under real-world load. The balance of motor power versus load reveals the real maximum speed. The intersection of the power curve (thick gray line) and the load curve (dotted line) occurs at 36 mph, which will be the real-world maximum speed. (Source: www.Ebikes.ca and www.kreuzotter.de.)

### Heat Dissipation in Motors

The maximum current that a motor can pass before it melts is determined by the heat transfer versus the resistive heating in the windings. The resistive heating is determined by the current and resistance of the windings. The motor heats up when the resistive heating power is more than the cooling power from heat dissipation. This is another power balance, and the result determines the temperature at which the motor will run at a certain current. If you know the melting point of the motor components, then you can use this balance to work out the maximum continuous current and power the motor can take. However, working out the heat transfer in a hub motor is much too complicated, and at present, there are no easy online simulators to do the work for you. Therefore, it is sufficient to know that you can increase the power of the motor if you also increase the cooling. The power normally is not limited by the motor anyway, but by the batteries (see Section 6.1.4 for overclocking and upgrading information).

### 4.1.3  Reviews of Different Hub Motors

Here are some short reviews of the different motors that are available. There are lots of other motors available, and the best place to get reviews is on the Endless-Sphere Forum at www.endless-sphere.com. Included here are the most common names on the market, but there are no-name and knock-off motors as well.

### Crystalyte Motors (X5, 400 Series, and BD)

Crystalyte motors are the oldest and some say they are still the best. Their power ratings are very conservatively rated for the power that you can put through them. The X5 is rated at 750 W continuous, but in reality, this motor can pump out 5 kW peak if desired. The 400 Series motors are rated at 500 W, but these can withstand 2.5 kW peak. Crystalyte brushed motors also can be overclocked way beyond their continuous ratings. Crystalyte motors are well built and don't break down. If you look inside them (see Section 5.4.4), there is nothing to go wrong; it is a very elegant design. The main disadvantage of the X5 is that it weighs a lot—12 kg/26.4 lb. The 400 Series motors weigh less and are easily sufficient for 99 percent of electric biking. The other disadvantage of the X5 is that it's difficult, but not impossible, to fit a disc brake on an electric bicycle with this motor. The axle spacing is too close and requires either very rare or custom-made parts. You can buy these motors and the Puma from Team Hybrid in the United Kingdom, ebikes.ca in Canada, or Island Earth in Australia.

### BMC (Puma)

The Puma motor weighs less and pulls more than the Crystalyte X5 but is not as reliable. The key to this is that it is geared to allow the motor to run at optimal speed while the bicycle runs at regular bicycle speeds. To achieve this, it has nylon gears that will fail eventually. The nylon gears can be replaced by steel gears, which are more reliable but heavier. The BMC comes in two versions, the V1 (Puma), which is 400 W, and the V2, which is 600 W. These are ridiculously low factory ratings given what the motors actually can pull. Both motors weigh about the same—4 kg/8.5 lb. The motor can accommodate a disc brake and a six-speed gear cassette and still fit into most bicycles.

### Tongxin

The Tongxin motor (Figure 4.4) is a small, lightweight, high-torque hub motor for riders who want a little assist with their bicycles on hills. The power is 180 to 250 W, and the weight is only 2.3 kg/5 lb. The motor is internally geared at a 12:1 ratio for climbing hills. Unfortunately, as of 2009, the motors have a high failure rate. Many people have experienced problems, and it appears that there are design flaws. The metal internal gears break, lots of controllers don't work with it, the motor is complex and difficult to repair, and it has a split-axle design that means that bumps cause it to skip. (Source: www.endless-sphere.com/forums/viewtopic.php?f=2&t=340&p=147452&hilit=tongxin#p147452.)

**FIGURE 4.4**  A Tongxin hub motor being used in a novel way.

### 9 Continents

The 9 Continents motor (Figure 4.5) is an internally geared hub motor. It has a larger diameter but is narrower than the Crystalyte 400 Series. It has higher torque and higher peak efficiency than the 400 Series but weighs 20 percent less. It is a good example of a middle- to low-price motor that's still fast enough. Currently, these motors are available with two winding types, a high-rpm Model 2806 for bicycles with 20-in wheels and a slightly slower Model 2807 for bicycles with 26-in wheels. For faster speeds, the 2806 motor can be laced into a larger wheel. The motors are 500 W continuous and 1.5 kW peak.

**20" 2806 Front Hub Motor:**                    *www.ebikes.ca*

Rim: 20 inches, 24mm wide
Weight: 6.1 kg
Motor KV: 10.3 rpm/V

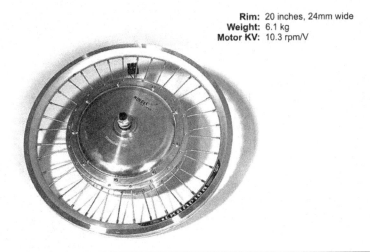

**FIGURE 4.5**   A 9 Continents motor.

### Golden Motors

Golden Motors produces 1,000-, 500-, and 200-w motors (Figure 4.6). The direct-drive motors seem to be solidly built. However, a number of people have encountered problems with the Golden Motors Company, as noted on the "Manufacturers to Avoid" page at www.endless-sphere.com/forums/viewtopic.php?f=3&t=236&start=60.

### Bafang

The Bafang motor (Figure 4.7) is a capable, high-torque motor that weighs only 3 kg/6.6 lb. It can come with disc brake mount on the front or back. The gears are plastic and can melt if it's used for serious power, but metal gears are available, so it is advisable to upgrade to these. The SWXH is the

**FIGURE 4.6**  A Golden motor.

**FIGURE 4.7**  A Bafang motor.

censored, cable-through axle, disc mount version. The Bafang is capable of 1,500 W of peak power.

### 4.1.4 Interfacing Motors with Controllers

Brushless and brushed motor controllers are not interchangeable, but you sometimes can swap controllers between different brands of motors if they are the same type. Most hub motors look the same, and usually, the only way you can determine the type of motor is by looking at how many wires it has coming out of it. With a brushed motor, you have only two wires, positive and negative, that can be swapped round for reverse movement.

With a brushless motor, you will have up to eight wires (Figure 4.8). The additional wires in the brushless motor are for the Hall sensor (positive, negative, and three sense wires) and because you have three coils not just one, so three thick power wires instead of two. The motor has three coils instead of one to even out the forces involved and increase efficiency. The Hall sensors in the motor are powered by a small voltage and switch their sensor output between high and low depending on whether they are over a magnet or not. The controller then picks up this positional information and the signal coming from the hand throttle to determine what power to give to the motor's three power wires (also known as *phase wires*). The power and sensor wires are linked through the controller and so must be connected to the controller in a specific combination.

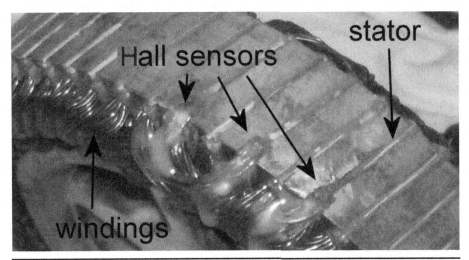

**FIGURE 4.8**  The inner workings of the Crystalyte X5 motor showing Hall sensors, stator, and windings. Each Hall sensor has three pins: positive (red), negative (black), and sense (blue, green, or yellow). The motor has eight wires coming out: three sense wires from the three Hall sensors, positive and negative for the Hall sensors, and three power wires.

Usually, if you buy a motor and controller together, the wires will be labeled or color-coded in an obvious way to make connection easy, but if you mix and match different brands of motor and controller, then you may need to contact the manufacturer or do some research on the Internet to find out the color-coding. You can use trial and error to find the correct wiring, but there are 36 combinations. Be gentle with the throttle if you are unsure about the wire combinations. You will know if the combination is wrong because the motor won't spin properly. It may oscillate or move only in a specific orientation. See Section 4.2.1 for more detail on how to find correct wiring combinations.

Some brushless motors have only three wires, with no Hall sensor wires. In this case, a special controller is used that employs the back–electromotive force (EMF) from the three phase wires to calculate the position of the motor. These controllers have more advanced electronics and can be more expensive. In addition, the motor needs to be moving for the back-EMF signals to be transmitted, so these special controllers are called *pedal-first controllers* because you can't use them from a complete stop (you have to pedal up to about 5 mph first before power can be applied to the motor).

### 4.1.5  Reliability of Motors

The reliability of the two types of motors is about the same. A brushed motor has brushes that gradually wear down and have to be replaced, but brushless motors have more wires and Hall sensors, which tend to break. The brushed motor is the simplest because it requires no complex electronics. The brushed motor needs only a controller to alter the speed, which is done by controlling the voltage. If a controller breaks on a brushed motor, it is possible to simply rig a switch on the handlebar to control the speed by on/off control. An emergency on/off switch is recommended anyway because when a brushed motor controller breaks, it often breaks in the full-power position, sending you flying with no hope of stopping!

### 4.1.6  Motor Types: Hub Motors

Motors can come in several styles of drive train, the way they apply power to the wheel. The main three types are hub motors, chain-drive motors, and friction-drive motors. Most electric bicycle motors on the market now are hub motors because of their simplicity of use. For this reason, this book will focus on hub motors. Hub motors are very easy to install and are very reliable. They come in front and rear hub wheel designs and may be sold installed in a wheel already or be supplied separately, requiring them to be laced into a wheel (see "Tips on Wheel-Building Hub Motors" in Section 3.4.15). Hub motors come in lots of different sizes, and usually the physical size relates to the power of the motor. Big, powerful motors usually have heavier-gauge spokes because this is needed for carrying powerful and heavy battery packs.

If you are going to push your motor to its limits, you should consider buying heavier-gauge spokes for it and building it into a stronger rim. You may have to do this anyway if spokes start to break or the rim gets damaged. If you are buying a motor from a retailer, you usually will have some choice between speed and torque variants of the same power rating. You also will be able to choose the wheel size for the motor. Choose carefully based on your needs. The motor geared for "fastest speed" is not always the best. A higher-toque motor geared for everyday speeds will have greater efficiency at lower speeds and better acceleration. Torque variants of a motor would be good for circuit racing with lots of corners, whereas a speed variant would be better for high-speed commuting or long, straight drag racing.

The torque versus speed efficiency tradeoff for a motor can be affected by the motor windings, battery voltage, and wheel size and gearing. For instance, if you want to make an electric bicycle that has so much torque that it does wheelies, then you would want a motor with a high-torque winding laced into the smallest wheel with a high-ampere controller. If you want to break bicycle land speed records, then a large-diameter wheel with a high-voltage motor wound for efficiency at high speed is best. High-power-rated motor, controller, and batteries obviously are essential in both the preceding designs.

Most decent hub motors now have a disc brake option. It can be a difficult job to modify a non–disc brake motor to accept a disc brake, but you will want disc brakes if you plan to go faster than 20 mph on a regular basis. So make sure that your motor can accommodate a disc brake.

### Hub Motor Gearing and Free Wheels

Hub motors also can be geared by internal gearing (Figure 4.9). Gearing allows a hub motor to run at higher rpms, which is more efficient for the motor, whereas the output is geared down to allow optimal speed for the electric bicycle. Some new hub motor designs allow the rider to change motor gearing while riding, but these don't exist in kit form at the moment.

Gears add complexity to the hub motor design and can be a point of failure. Some gears are made of nylon, and these are more likely to fail than metal gears but are quieter and lighter in weight. Some hub motors are free wheeling and some are fixed. *Free wheeling* refers to what happens when the power is taken away in a coast. If the motor has a free wheel, then it will coast longer when the throttle is released. If it has no free wheel, then the motor will try to regenerate when the throttle is released. Some controllers have regenerative braking capability and can be used with non-free-wheeled motors to recover braking energy back into the batteries and extend the

**FIGURE 4.9**   Inside a Puma geared hub motor.

range of the electric bicycle. Most controllers currently don't have regenerative capability, and in this case, a non-free-wheeled motor will be less efficient while coasting.

### Inside a Hub Motor

Hub motors have powerful neodymium-iron-boron magnets around the inside of the hub wheel, so be careful when taking them apart. These magnets attract and stick to the stator axle part of the wheel. This force is very strong, and the attraction is violent. When you put the stator axle back into the magnet rim hub, you will feel the attraction as it goes in, and then it will jump suddenly as the force of the magnets takes over. When you put these parts together, be careful that you don't get your fingers trapped in the gap.

Often the spokes on a hub motor are under high tension, and this can distort the case off the hub motor. When you take both side plates off a hub motor, sometimes the spokes will stretch the casing of the motor so that is difficult or impossible to replace the side plates. If this happens, the spokes must be loosened before the side plates are screwed on, and then the spokes can be retightened. This is the case with the Crystalyte X5 hub motor. Be careful not to let water into your motor because the magnetic and electrical action inside can cause accelerated rusting that will damage the insides of the motor.

From time to time, you will have to replace the bearings in your hub motor. The two bearings in the side plates support all the weight of the bicycle and so will wear quite fast. With daily use, you may need to change the bearings on a heavy bicycle once a year. If you have the hub open for any reason, it is a good opportunity to do some preventive maintenance and change the bearings. If a bearing fails while in use, you will instantly notice bad noises coming from the motor and a noticeable loss of smoothness. If a bearing failure occurs, stop using the bicycle because continued use could damage the motor.

You can test if the bearings are broken by holding the bicycle by the saddle and shaking it back and forth sideways. If the motor pivots on the axle, then the bearings are shot. The air gap (between the magnets and stator coil) usually is very small, and with bad bearings, the magnets could scrape against the stator, which could damage them or the Hall sensors.

### 4.1.7 Motor Types: Chain-Drive and Friction-Drive Motors

Chain-drive setups (Figure 4.10) require a bit more effort to put in place than hub motors but have several advantages. With chain-drive motors, it is possible to change the gearing, which is not possible on many hub motors. This will increase the motor's efficiency. It is possible to use the main bicycle chain and allow the user to change the gearing that the motor uses while riding. Chain-drive motors are smaller than hub motors for the same power rating. In addition, chain-drive motors can be suspended using the bicycle's suspension, unlike hub motors, so this reduces the bicycle's unsprung weight. Because the motor is outside the wheel, it should be cheaper and easier to replace the wheel if it is damaged. It is for this reason that most larger electric vehicles (EVs) don't have hub motors and many are chain-

**Figure 4.10**    Chain-drive setup. Several different motor mounting positions can be used, and the chain can drive the main sprocket and an additional rear gear wheel.

driven. However, the mechanisms are often more mechanically complex and difficult to install initially.

Usually, chain-drive kits include some method of mounting the motor, but there are many different types of frames, which can make connecting chain-drive motors difficult. There is no standard motor mount, unlike hub motors, which are standard and fit all bicycles. If you just buy the motor, then it can be quite technically challenging to fix it to the bicycle. Often purpose-built chain-driven electric bicycles will have a motor mount already welded to the frame, but even with this, it still can be challenging to change the motor if an upgrade is desired. The difficulty comes from having to correctly tension two chains at once.

Friction-drive systems (Figure 4.11) now are regarded mainly as an obsolete technology because of problems with slippage during wet conditions, special tire requirements, and inefficiency. Friction-drive systems have much the same problems as rim brakes. A friction-drive system can work well in certain circumstances, though, and it provides an easy way to gear down an electric motor for driving a bicycle. You also can get belt-driven drive trains, but these are more common on motored bicycle engines, where you can't easily build the engine into a wheel hub.

**FIGURE 4.11**    Friction-drive system using the rim. It is also possible to use the tire surface instead.

## 4.2 Controllers

With modern brushless motors, the controller is the heart and mind of the machine. However, the controller is a complicated beast that can be treated by the user as just a magic black box—and as long as it works, that's often sufficient. Therefore, this section mainly addresses controller theory that a novice electric biker might want to skip.

Modern controllers have many functions, including controlling power, reversing, and recovering braking energy when stopping. Some controllers can interface with other devices such as advanced electric bicycle computers or data loggers. Often a basic low-power controller and a faster high-power controller will be built in the same factory and have the same chip set and circuit board. Sometimes lower-power controllers can be modified to achieve more power just by upgrading a few key components or using simple modifications. Often certain features are present on the circuit board that users don't know about. Sometimes these controllers can be modified to increase the level of control that the user can enjoy. The aim of this section is to make you aware of these options so that you can be more in control of your electric bicycle's performance, battery health, and functionality.

### 4.2.1 Finding the Right Wiring Configuration for a New Motor-Controller Combination

Most of the time, brushed controllers (Figure 4.12) are interchangeable between brushed motors, assuming that the rating specs are obeyed. Brushless controllers, however, are usually not easily interchangeable between brushless motors because there is no standard color code for wiring the phase wires and Hall sensor wires. Also, the Hall sensors can be positioned in two different ways in relation to the magnet poles. Therefore, to fit a brushless controller to a different motor, some experimentation is needed unless you already know the combination.

1. You can use trial and error to find the correct wiring, but there are 36 combinations.
2. Fit the phase wires and Hall sensors with plugs or a terminal block so that you can make and change connections easily.
3. Pick a combination of controller to motor phase wires and controller to motor sensor wires. The phase wires are always of much larger thickness than the sensor wires. Always connect thick to thick and thin to thin.
4. Connect the wires, and then gently turn the throttle.
5. You will know that the combination is wrong if the motor won't spin properly. It may jump, oscillate, or move in only one specific orientation.

**Figure 4.12** Brushed motor controller. This controller is sealed with potting compound and is difficult to disassemble if repairs are needed.

6.  Keep working through the combinations until you find the right one. You can protect the controller from damage during this trial-and-error process by putting a light bulb in series between the battery and controller. A MR16 bulb often will fit easily into the fuse holder.

On some motors, there may be a further complication that the position of the Hall sensors relative to magnet poles is different from what the controller was designed for. Some controllers have a jumper that can be used to specify what type of configuration is used, 60 degrees or 120 degrees. The best way to connect a motor or controller is to look on an Internet forum to see if the wiring for that motor-controller combination is known. See "Known to Work Brushless Motor/Controller Wiring Diagrams" on www.endless-sphere.com/forums.

The Hall sensors signal when the power pulses should be sent. Therefore, the Hall sensors have to be positioned in a specific orientation for the controller or the timing of the power pulse will be off. There are two types of windings: delta and wye configurations. Some controllers have a jumper switch that allows for switching between either type of motor winding configuration. If the wrong winding configuration is selected, then the motor will not have any power, and the controller may be damaged. There may be one wiring combination for forward and one wiring combination for backward movement. Some controllers also have a reverse switch, and the speed of the motor may be limited when in reverse mode.

### 4.2.2 Inside a Controller

One way to destroy a controller is to place the motor under high load when it is not rotating and then set maximum throttle. This will cause a very high current to flow because it is basically like shorting out the batteries. The capacitors in the controller will receive a massive power spike and blow up. This will protect the motor, which would be next to blow if the controller didn't. The windings would melt under the extreme high current owing to the lack of back-EMF because the motor is not moving. This failure usually can be repaired simply by replacing the capacitors. If the capacitors are too badly exploded to make out their value, then you can either search online or guess or use the specs from Section 5.4.1. The voltage spec is likely to be at least 1.5 times the maximum voltage of the controller, so just get the highest-value electrolytic capacitor that will fit in the case. Pay attention to polarity when soldering it in place (see Section 5.4.3 for help).

Most controllers have an onboard thermal cutout circuit that will protect the controller from overheating. If the controller overheats, it will shut down and will start again only after it has cooled down. It's best to position the controller in a location with airflow (not inside a bag) so that this will not happen.

All controllers also have a low-voltage cutout circuit (LVC). This is there to protect the batteries but is a very crude form of protection. Controllers usually will accept a range of different currents, but the LVC is a set value that will not change to match the type of battery or voltage used. The LVC is designed to stop the controller if the battery voltage drops below the LVC value. The LVC is specific to the voltage of the batteries you are using. This is why some controllers have voltage stickers on them when you can easily exceed the stated voltage. You only have LVC protection when you use batteries of the stated voltage. This may occur if the batteries are either empty or being discharged too fast above their C rating. The circuit is usually a comparator type of circuit with resistors and a diode. The LVC usually can be altered by altering the values of the resistors in the LVC.

The controller measures and limits current by measuring the voltage drop across a current shunt resistor (Figure 4.13). Based on this measurement, there is a feedback mechanism that will either increase or reduce the pulse-width modulation (PWM) of the controller based on the throttle position. On any controller circuit board, the current shunt resistor usually will look like a few parallel bars of wire. To raise or lower the current limit of the

**FIGURE 4.13**   Shunt resistor and main capacitor.

controller, it is possible to either add solder or cut the shunt to alter the re-sistance. This tricks the controller into thinking that there is either less or more current flowing through the circuit than is actually the case and changes the limit accordingly. The feedback mechanism uses a potential divider circuit with a shunt resistor. It is possible to change the values of the other resistors in the potential divider circuit to alter the current limit. A potentiometer can be installed to have a variable current limit. See Section 6.1.7 for variable-current-limit modifications.

## 4.3  Batteries

The current rate of technological advancement in the field of batteries is truly astounding. There are new battery breakthrough news stories about once a month, and new products are introduced every few months. Over the last 5 years while I have been involved in electric bicycles, I have watched electric bicycles move away from heavy lead batteries, through nickel–metal hydride (NiMh) and nickel-cadmium (NiCd) batteries, to advanced lithium polymer and lithium-iron-phosphate batteries. These advances have moved electric bicycles from simple toys into genuine versatile transport solutions. Many parallels can be drawn with the world of remote control (RC) hobbies, where these same battery advances have made it possible for electric RC models to beat gas- and nitro-powered RC engines in terms of shorter lap times and higher power-to-weight ratios. Surely as battery EVs improve, so will passenger vehicles start to follow.

### 4.3.1 Battery Theory

A *battery* is a collection of cells working together, and a *cell* is a discrete element of a battery. In popular language, these terms are often used interchangeably, but not in this book. A battery is a means to store energy by way of chemical reaction. This reaction may be reversible, as in rechargeable batteries (secondary cells), or irreversible, as in "throw-away batteries" (primary cells). Here we are concerned only with secondary cells for electric bicycles—for obvious reasons!

A cell normally consists of electrodes, electrolyte, a separator material, current collects, contact tabs, a casing, and insulation. The energy in a cell is stored by chemicals on the electrodes. The electrolyte allows the transport of ions between electrodes. The separator is insulation material that separates the electrodes but allows them to be close together for conduction of ions. The insulation stops the battery from short circuiting, and the casing keeps everything together.

When you discharge a cell, you allow a path for electrons to travel from one electrode to the other to complete the chemical reaction. When you recharge a cell, you reverse the flow of current and force the reaction back the other way. If you mistreat a battery by overcharging, overdischarging, or short circuiting, other bad chemical reactions can happen that are irreversible because of loss of chemicals or thermal damage.

The main parameter of a battery is the energy density, which can be measured in watthours per kilogram/pound. This is a measure of how much energy is stored inside the battery compared with its weight. The maximum energy density of each battery chemistry can be predicted by theoretical equations of the chemical reactions occurring in the cell. However, the theory is based on the active materials present, and real batteries never reach the theoretical limit because they need other auxiliary components such as insulation and casings to support the battery, and these add weight. Real batteries rarely reach half their maximum theoretical energy density.

#### Internal Resistance of a Cell

A cell is a source of power, but it also contains resistance because the chemical reactions inside the cell can occur only so fast. This is usually limited by the transport of ions through the electrolyte. This resistance is called a cell's *internal resistance*. Good cells have a low internal resistance. This means that you can draw current quickly from them. Internal resistance cannot be measured by a multimeter because it is based on the physicochemical properties of the cell. When you draw current from a cell, its voltage will drop because it loses power to its internal resistance. This is called *voltage sag*.

Voltage also can be dropped across the battery tabs. This energy goes into heating the cell and is lost. Therefore, there is always a limit to how fast you can discharge a cell before it either melts or does other damage to itself. With larger battery packs, the self-heating is increased because there is a smaller surface-area-to-volume ratio, which makes the heat dissipation worse. You can calculate a cell's internal resistance by using Equation (4.1). This equation also can be used to compare the quality of different cells. Internal resistance will change as the cell gets older and as its state of charge changes. You can test cell resistance with an ESR meter, which is designed to measure the equivalent series resistance (ESR) in capacitors. Some ESR meters require the connection of a capacitor in series to prevent automatic discharge of the battery.

---

**EQUATION 4.1**    Calculation for cells internal resistance.

$$R_b = \left(\frac{V_b}{I_L}\right) - R_L$$

### Difference Between Capacitors and Batteries

A battery can be optimized for either high capacity or high discharge rate. A high-capacity battery will have large electrodes or thick layers of coating on its current collectors, whereas a high-discharge battery will have electrodes with high surface area. Capacitors called *supercapacitors* can be used as another source of power for EVs. The difference between capacitors and batteries is that capacitors store their power only along the surface of their electrodes, whereas batteries store their power in the bulk chemicals contained within the battery. Because a capacitor does not store energy in its bulk, the surface area can be maximized for high power discharge rates. Capacitors are not limited by transport of ions through the electrolyte, so they have a much lower internal resistance. However, because a capacitor does not store power in its bulk, this means that it has only limited energy storage capacity compared with batteries.

### Cell Voltage Effects

A cell's voltage is determined by the electrode reactions that it supports and can be predicted for any electrode pair by theoretical equations. The voltage of a cell is normally between 0 and 4 V. The voltage goes up when the cell is being charged and down when it is being discharged. You can use the resting voltage or *open-circuit voltage* as a measure of how much charge is left in the battery. This can be difficult, however, because a good battery will have a flat discharge curve (voltage-versus-state-of-charge graph). Each cell chemistry has a nominal voltage, a fully charged voltage, and an "empty" voltage. It is important never to discharge any cell past its empty voltage because it will damage the cell. Never let any cell reach 0 V. It is important never to charge a cell past its fully charged voltage because this also will damage the cell. The nominal voltage is the cell's working voltage. During discharge at the cell's specified discharge rate, the cell will be around this nominal voltage. If the discharge is faster than the cell's specified discharge rate, the voltage will sag to a lower voltage. If the discharge is slower than the cell's specified discharge rate, the voltage will be higher, closer to the cell's fully charged voltage. There is no hard and fast limit between damagingly fast discharge and the normal discharge rate. The difference is more of a tradeoff between cell life expectancy and discharge rate. This is why cells sometimes have three specified discharge rates: a peak rate, a normal rate, and a battery management system (BMS) cutoff discharge rate.

In addition to energy density and discharge rate, there are several other factors that can affect batteries that are specific issues related to their chemistry. Charge/discharge profile, charge/discharge endpoint determination, overcharging/overdischarging failure mechanisms and protection mechanisms, and series/parallel behaviors are all related specifically to the battery chemistry and need to be looked at individually. Often, high-tech battery chargers and BMSs are needed to manage the charging and/or discharging of a battery to prevent it from going outside its safe operating limits.

### 4.3.2  Pack Building and Series/Parallel Effects

Large battery packs are often made of combinations of series- and/or parallel-connected cells. When cells are connected in series, the current that passes through each cell is always the same. When cells are connected in parallel, the voltage across them is always the same. The ideal battery pack would just consist of one cell. Multiple-cell battery packs usually need to be balanced, and more cells increase the likelihood of having a bad cell. However, one cell does not have a high enough voltage for an EV, so several cells need to be connected in series to increase the voltage; otherwise, the current requirements would be too high. Therefore, the ideal EV battery would consist of several very large cells connected in series to reach the desired voltage. Often we have to connect cells in parallel to increase pack capacity. Cells are manufactured in discrete capacities, and larger-capacity cells may not be available. The parallel connections can be made before or after the series connections. That is, the battery back can consist of parallel-connected cell groups that are then series connected, or it can consist of series-connected strings that are then parallel connected with other battery packs. There are advantages and disadvantages to both.

## Cell Balancing in Series-Connected Batteries

Series- and parallel-connected cells have different balancing characteristics. When a battery pack is assembled, the manufacturer should test and select cells that are closely matched in terms of capacity and internal resistance. If a battery pack is not balanced when it is charged or discharged, then bad things will happen. A balanced pack is one where all the cells have the same state of charge and capacity. If a series-connected battery pack is unbalanced during charging, then some of the cells will reach full charge before others because the same charging current runs through them all. The cells that reach full charge first then will be overcharged if the charging continues. If a series-connected pack is unbalanced during discharge, then some of the cells will be completely discharged before the rest, and if discharging continues, the empty cells will be forced to discharge further by the other cells and will become damaged by voltage reversal. It is clear that when in series, cells always should be the same capacity. It is often said that "a battery pack is only as good as its weakest cell."

## Parallel-Connected Cells

When cells are connected in parallel, their voltage will be the same. If two cells of different voltages are parallel connected together, then their voltages will both change, and the resulting voltage will be somewhere between the starting voltages of the two cells. If the two cells were of the same type and just of different states of charge, then the higher-charged cell will just charge the other cell until they are at roughly the same state of charge. The charging/discharging may occur very fast, though, and this may damage the cells depending on the difference in state of charge and the C rating of the cells. If two different types of cells with different nominal voltages are connected in parallel, then bad things certainly will happen. The power from the higher-voltage cell will overcharge the lower-voltage cell, potentially causing damage. At the same time, the load on the lower-voltage cell may overdischarge the higher-voltage cell, potentially causing damage. There is an exception to this if diodes are used to prevent cross-charging between the cells (Figure 4.14).

## Series/Parallel Cell Failures Inside Batteries

When a battery fails, it can fail either open circuit or short circuit, or its voltage could fall outside its normal parameters and fail to produce current. Different series/parallel configurations will react differently to the different types of cell failure. If a series string of cells has a single cell that fails open circuit, then this will stop the whole string from working (Figure 4.15) until the dead cell is replaced. It will read 0 V if a voltage meter is put across it. If a series string has a cell that fails short circuit, then the battery will continue

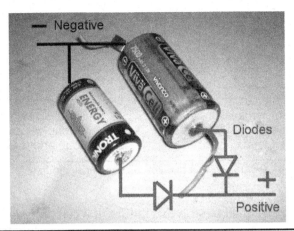

**FIGURE 4.14**  Use diodes to protect cells of different types or different voltages when connected in parallel. This diagram shows how to connect the diodes. These batteries are actually the same voltage and type and thus would not need diodes.

to work but at a reduced voltage. If this series string is in parallel with other similar packs, then the string will get charged by the other packs and possibly overcharged. If a series string has a cell that fails with its voltage drop outside normal operating parameters (i.e., there is a "weak cell"), then it should be replaced because this will reduce pack performance. Continued use of the pack in this condition will result in the cell failing even further and possibly failing open circuit.

With parallel cell groups, the effects of failure are different. A cell failing open circuit in a parallel group will reduce the capacity of that cell group but still allow continued operation (Figure 4.16). If this cell group is part of a larger series string of cell groups in a battery pack, then the capacity of the whole battery pack will be reduced to the level of the bad cell group until the cell is replaced (like a "weak cell" in a series string). Continued use of the unbalanced pack will result in the rest of the cells in the parallel group being damaged by overcharges and overdischarges until the whole cell group dies.

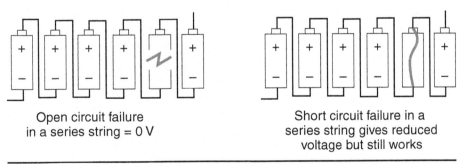

Open circuit failure
in a series string = 0 V

Short circuit failure in a
series string gives reduced
voltage but still works

**FIGURE 4.15**  Cell failure in a series-connected string of cells.

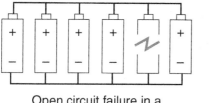

Open circuit failure in a
parallel cell group gives
reduced capacity

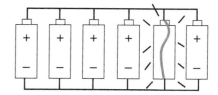

Short circuit failure in a parallel cell
group will short circuit the whole
group and could damage other cells

**FIGURE 4.16** Cell failure in a parallel-connected cell group.

A BMS is designed to stop this failure. If a cell fails short circuit in a parallel cell group, it will short out all the other cells in the parallel group, which can create a lot of heat. Depending on the type of battery chemistry, the heat from short-circuited failed cells can damage other neighboring cells. If this damage is sufficient to cause internal short circuits in the neighboring cells, then it can cause the pack to go into thermal runaway in a spectacular fashion, but this is rare.

### 4.3.3 Battery Safeguards

There are lots of things you can do to protect your batteries and keep them within their safe operating limits. I have already talked about fuses (Figure 4.17), and they are essential in any battery pack to protect the batteries and wiring from going into meltdown because of a short circuit. Thermal fuses (Figure 4.18) also can be used to protect a battery pack. These are like normal fuses, but they blow when a high temperature is reached instead of a high current. These are often seen inside nickel batteries, which get hot dur-

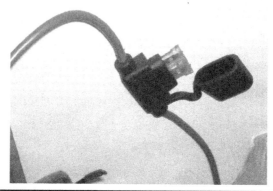

**FIGURE 4.17** Automotive fuse holder.

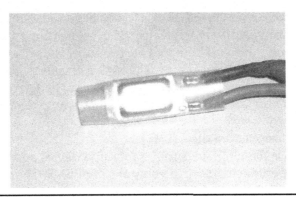

**FIGURE 4.18**    Thermal fuse.

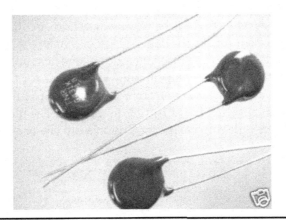

**FIGURE 4.19**    Thermistors.

ing use and at the end of charging. Thermistors (Figure 4.19) are resistors that change their resistance based on temperature. You see thermistors in NiMh battery packs because they are part of the charge-termination method for charging them.

Because nickel batteries don't follow a constant-current/constant-voltage charging pattern, they have to use a special method to detect when they are full. Nickel batteries for electric bicycles will have a three-pin plug connecting the charger to the battery. Two of the pins are obviously positive and negative, but the third goes to a thermistor and then to negative. The special charger measures the resistance of the thermistor as the battery is charged, and when the battery gets hot at the end of charge, the resistance reduces, and the charger notices this and stops charging. Batteries can be isolated from each other to prevent them from cross-charging by using diodes (see Figure 4.14). The diodes will only allow charge to flow in one direction; this has the advantage that you can use different batteries in parallel with each other that otherwise would have had damaging effects because of their voltage differences causing cross-charging. Low-voltage cutouts (LVCs) are a safeguard built into the controller to ensure that the batteries are not drained completely empty. The LVC is usually a crude method, because most controllers will accept a range of different battery voltages, but the controller will cut out at only one voltage that is not selectable.

### Battery Management System (BMS)

The BMS is the ultimate battery protection device and has many functions, much like a little hard-wired computer. The BMS (Figure 4.20) is always attached to the battery and forms part of the battery pack but draws very little power from the pack. The BMS will limit the discharge rate and cut out if the rate exceeds the safe limit of the battery. Sometimes the BMS cutout is faster than a fuse at protecting the battery from a short circuit. The BMS monitors all the cells or cell groups in the battery pack via balance wires to monitor the voltages of the cells. If during discharge the voltage of a cell group drops below the safe limit, then the BMS will cut out. When charging, the BMS will "balance" the pack by draining charge from cells that have reached the desired voltage already, thus allowing the other cells to catch up. The BMS uses resistors to discharge the cells during balancing, and this can create a bit of heat.

**Figure 4.20**    16 cell battery management system (BMS) for a 48V LiFePo$_4$ battery.

## 4.3.4  Commercially Available Batteries for Electric Bicycles

### Lead Acid (SLA, Pba)

This was the first battery chemistry available to EVs and has been around for over 100 years. These batteries are still used in EVs today but only very slow ones such as golf carts and neighborhood EVs. A single lead-acid cell is 2 V (nominal), so lead-acid batteries are always a multiple of this voltage (i.e., a 12-V battery is six lead-acid cells in series). The fully charged voltage is 14.5 V for a 12-V battery, and the empty voltage is 11 V. Lead-acid batteries come in two types: deep discharge (Figure 4.21) for EVs and normal for starting car engines. The difference is in the way the electrodes are made, with deep-discharge electrodes having larger plates that can hold more capacity, whereas normal lead-acid batteries have high-surface-area plates that can deliver at a faster rate. Flooded lead-acid batteries are also available. These batteries have openings where you can add distilled water or acid to keep the battery in good condition.

The main advantage of lead-acid batteries is that they have a high discharge rate, about 9C, and they are very cheap, usually two or three times cheaper than other, better batteries. Also, because they have been used in cars and are an established product that has been used for a long time, 95 percent of all lead-acid batteries are recycled.

However, they have several big disadvantages. These batteries have a very small energy density, only 35 Wh/kg or 16 Wh/lb, which means that any lead-acid battery–powered electric bicycle will have a very limited range and be very heavy. A lead-acid battery–powered electric bicycle will need 1 kg/2.2 lb of battery for each mile of range. They can only be recharged slowly, which limits their flexibility of use. They like to be kept charged all the time to maximize their life. Too much depth of discharge (DOD) will reduce their life. Their life cycle under electric bicycle conditions is only usually 250 cycles (Lemire-Elmore, 2004). Such a short life cycle often makes it more economical to buy a better battery chemistry that will work out cheaper in the long run. Lead-acid batteries should be charged using a con-

FIGURE 4.21   Lead-acid battery for a mobility scooter.

stant-current/ constant-voltage (CC/CV) charge profile, usually C/10 until 2.4 V per cell is reached, and then that voltage should be maintained while the current levels off. Lead-acid batteries can be charged in parallel and/or series with other lead-acid batteries without problems. No BMS is required.

### Lead-Acid Battery Safety

Lead-acid batteries contain 33 percent sulfuric acid as the electrolyte. The acid should never leave the battery because such batteries are sealed. If there is a rupture and the acid escapes, then be careful because it can burn the skin and destroy clothes. Wash with water if you suspect that you have battery acid on your skin. A small amount of baking soda in water can be used to neutralize the burn. Overcharging a lead-acid battery with high voltages will generate oxygen and hydrogen gas by electrolysis of water, possibly forming an explosive mix. This is why some renewable-energy battery banks are ventilated.

### Lead-Acid Battery Formats

There are several standard formats for cells and batteries that exist for other applications that have been adopted by bicycle manufacturers. For old lead-acid batteries, there are several standard sizes that were brought about by car, wheelchair, and golf cart manufactures. Lead-acid batteries come in square bricks of different shapes and sizes. The smaller batteries (<20 Ah) come with solder or crimping tabs, and the larger car batteries (>20 Ah) come with screw or clamping terminals.

### NiMh Batteries

Nickel–metal hydride batteries are a good all-round battery chemistry and a much better choice than lead-acid batteries for electric bicycles. These batteries are used in hybrid vehicles because of their fast recharge capability for regenerative braking and because they are reliable. NiMh batteries have a higher energy density than lead-acid batteries and can hold around 65 Wh/kg or 30 Wh/lb. This is slightly lower than most lithium batteries, but NiMh volumetric energy density is higher than that of many lithium batteries, so they are still a good choice for EVs.

NiMh cells can be charged at up to 1C but usually are charged more slowly for extended life (around C/5). The discharge rate is reasonable, although it varies among brands and is usually 1C to 5C. These batteries are roughly twice as expensive as lead-acid batteries, usually around 2 Wh/£ or 1.33 Wh/$ at the time of this writing, but they also last longer—300 cycles under electric bicycle conditions (Lemire-Elmore, 2004). NiMh batteries don't need to be kept charged all the time like lead batteries. The nominal working voltage of an NiMh cell is 1.2 V, with a full voltage of 1.4 V and an empty voltage of 1.0 V.

Charge termination is done by measuring temperature and looking for the temperature increase at the end of charging. Balancing is done by slowing the charge current at the end of charging. Full cells will harmlessly dissipate the slow charge current as heat. This function is built into NiMh chargers and is designed to work with a thermistor located inside the pack. A simple 40°C thermostat circuit also will work by switching the current off when the cells get hot and allowing them to cool off and then switching on/off, maintaining the temperature until all cells are balanced. No BMS is needed, and NiMh BMSs are not readily available. This is a shame because a BMS helps to prolong the life of the pack by preventing some forms of misuse by the user. The batteries are damaged by temperatures over 70°C. A temperature over 100°C will destroy them. NiMh cells and batteries cannot be fast charged in parallel without a basic BMS. Parallel fast charging can result in severe overcharging and failure of one of the parallel cells or batteries.

Discharge is different, however, and NiMh packs can happily be paralleled on discharge. Diodes or switches can be used to get around the problem of charging parallel batteries. This is highly recommended if you have parallel NiMh batteries. Some NiMh chargers sense end of charge by measuring a voltage peak. NiMh cells also can be "trickle charged" by a very low current when the heat caused by the small overcharging current is insignificant.

## Nickel Battery Formats

NiMh cells come in several standard sizes adapted for toy manufacturers and hobbyists, including AA (<3 Ah), AAA (<2 Ah), C (<5 Ah), D (<12 Ah), and F (<20 Ah). These cells use a cylindrical metal can design and can come with metal tabs welded onto them for connection to other cells. To make larger EV-sized battery packs, these cells are welded together and then packaged with protective packaging. The ones that are mainly of interest to electric bicycle enthusiasts are D cells (Figure 4.22), but C and F cells are also used. The disadvantage of cylindrical cells is that they leave lots of unused space between the cells when assembled into packs. For specialist applications such as EVs, there are "prismatic" cells, but they are often not available commercially to the public. Prismatic cells are large (20 to 500 Ah) cells that are rectangular in size and usually have screw terminals like car batteries. These cells are easier to build into a large EV pack; they have a lower cell count and no void space need be left between the cells. Prismatic cells are used in hybrid and EVs, but electric bicycle enthusiasts have gotten hold of wrecked hybrids and made packs out of the working cells.

## NiMh and NiCd Safety

NiMh and NiCd batteries use potassium hydroxide as the electrolyte, which is alkaline. If you severely overcharge an NiMh battery, it will boil the electrolyte,

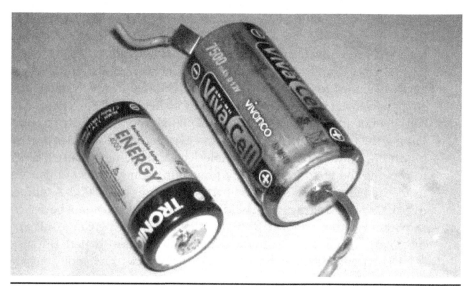

**FIGURE 4.22**   C cell (left) and D cell (right), both nickel–metal hydride cylindrical cells.

and the battery will vent. This will ruin the battery, and the fumes are toxic. Don't breath the fumes. Evacuate the area, and wait for the fumes to dissipate. Turn off the charger at the socket, and open windows to assist ventilation.

### NiCd Batteries

NiCd batteries are very similar to NiMh batteries, and all the preceding applies to NiCd cells as well as NiMh cells. Even the voltage is the same, and the same battery chargers will work with both chemistries (with some exceptions). The main differences are that NiCd batteries have slightly less energy density (35 Wh/kg or 16 Wh/lb) but a much longer life (1,000 cycles) (Lemire-Elmore, 2004). Many people believe that NiCd batteries have a "memory effect," where a half-discharged cell, once recharged, will discharge only down to the previous level of discharge. This effect seems to be an urban legend (Buchmann, 2001). The memory effect has been noticed only on a very specific application, a satellite, where the cells were consistently discharged down to exactly the same level on each cycle and over time built up a "memory." In electric bicycle use, this would never be the case, and as long as depth of discharge varies even slightly between rides, the memory effect will not occur. The same parallel charging problem occurs with paralleled NiCd cells as with paralleled NiMh cells.

### Lithium Polymer (Lipo, Lion, Lithium Colbalt) Batteries

Lithium polymer batteries are the battery of choice for RC hobbyists. These batteries have the highest energy density of any secondary cell (150 Wh/kg

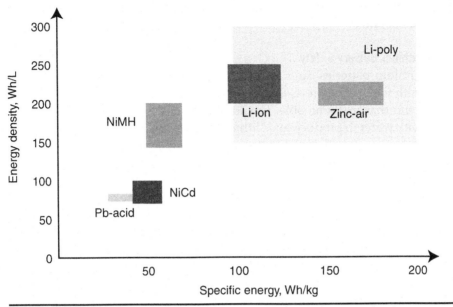

**FIGURE 4.23**   Graph of volumetric energy density versus mass energy density.

or 68 Wh/lb; Figure 4.23). These batteries also can reach extremely high discharge rates (30C) and can be fairly cheap. The only disadvantage is that that they are unstable, and if abused, they will explode in a dangerous fireball. Several house fires have been caused by lithium polymer batteries, and it is advised that these batteries be recharged outside the house in a non-flammable area. This is in case you have a charger malfunction that causes a battery explosion.

Several laptop manufacturers had to recall their batteries because of spontaneous fires caused by this type of battery. No large manufacturer has yet used them in EVs because of this problem. Large EV packs often contain hundreds of cells, which increases the chance of failure. The high temperatures caused by some cell failures can cause other neighboring cells to fail, which, in theory, could cause a chain reaction that would destroy the EV. As long as none of the cells is overcharged, overdischarged, physically damaged, or overheated, then these batteries should be fine, and many RC hobbyists have had many happy cycles from them.

Lipo cells have a nominal voltage of 3.7 V, a full voltage of 4.1 V, and an empty voltage of 2.5 to 2.9 V depending on type. Expensive specialist charging and BMSs are required, and the cells never must be allowed to exceed the full voltage of 4.1 V per cell. Several companies are currently working on safer lithium polymer batteries that can be used in EVs, which is now seen as the biggest untapped marketplace for batteries. The main effort seems to be replacing the cobalt with other elements such as manganese or phosphate.

## Lithium Battery Safety

Lithium batteries can't use water-based electrolytes because lithium reacts with water. Therefore, the electrolytes used in lithium cells are usually solvents that are flammable. The lithium is highly reactive and will react quickly with water to produce toxic lithium hydroxide. Lithium also will react slowly with oxygen in the air, but the reaction will be sped up if moisture is present in the air. Very rarely, some lithium polymer batteries (used in laptops) have cells that contain manufacturing defects. These defects may cause them to combust spontaneously for no reason. Poor Chinese quality control is usually to blame for these faults. Several laptops and mobile phones have had to be recalled for this reason. If you don't believe me, watch the videos on YouTube of people hammering nails through them! If you are new to EVs or electric bicycles, then don't use lithium polymer batteries. Many people just don't use lithium polymer batteries at all because of the fire risks. Lithium–iron phosphate (LiFePO$_4$) batteries are a much safer alternative.

The chemicals in most batteries are toxic to some extent. If there is a battery fire or venting of electrolyte, then do not breathe the toxic fumes. Get away as fast as possible, and call the fire department if necessary. Disconnect any smoking battery at the main circuit if it is plugged into a charger, but do so only if safe. If the battery is smoking, move it outside if it is safe to do so. If the battery is in a safe location outside, then just let it smolder. If the battery is on fire inside, then extinguish the fire with a fire blanket and a $CO_2$ extinguisher, but only if it is safe to do so. It may not be possible to extinguish the fire because of the chemicals involved. Do not use water fire extinguishers. Lithium reacts with water. Wear protective equipment when extinguishing a battery fire in case the battery explodes or bits are blown around by the fire extinguisher.

The highest risk is when you are working on an open battery. When working on batteries, avoid using metal tools whenever possible; work on a clean, empty, nonconductive surface; and don't wear jewelry. If a cell is broken or damaged, do not attempt to fix it. It is impossible anyway because individual cells were not made to be user serviceable, and they are very complicated.

## Lithium Battery Formats

Lithium cells don't have any standard format. Some manufacturers use the 18650 cell (Figure 4.24), which is a metal can $18 \times 65$ mm/$0.7 \times 2.6$ in. This is the format used by A123 systems. Many lithium cell manufacturers use a foil pouch for their cells (Figure 4.25) because it is lightweight and offers a very good energy density, but the disadvantage is that it offers little me-

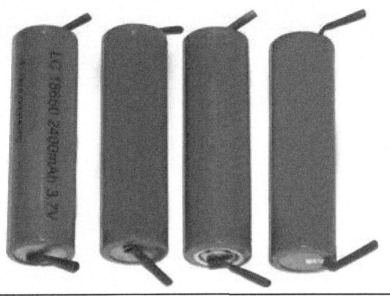

**FIGURE 4.24**   Cylindrical metal cells of the 18650 type with tabs.

**FIGURE 4.25**   Foil pouch cell, lithium–iron phosphate (LiFePO$_4$).

chanical protection to the cells, and they have to be protected inside a padded strongbox. The largest cells are prismatic cells (Figure 4.26), which use a larger plastic case to contain a single cell. These are for electric cars and motorcycles.

**Figure 4.26**    Large prismatic cell for an EV.

### Lithium–Iron Phosphate (LiFePO$_4$) Batteries

Lithium–iron phosphate batteries have recently become the battery of choice for electric bicycles. These batteries have a high energy density (100 Wh/kg or 45 Wh/lb, 200 Wh/L or 200 Wh/qt) and can reach extremely high discharge rates (up to 30C). Many different manufacturers exist, and their products vary considerably, especially in terms of price and discharge rate (1C to 30C). Charge rate can be up to 1C but is usually lower, around C/5, to ensure longer life.

LiFePO$_4$ is accepted as a safe chemistry; it won't explode if abused. If it does fail, it is more likely to balloon if it is a foil pouch type of cell. LiFePO$_4$ cells are charged using a constant-current/constant-voltage charge profile (CC/CV). This usually means charging at C/5 until 3.75 V per cell is reached and then maintaining this as the current levels off. LiFePO$_4$ cells tend to drift out of balance with other cells in the same pack and therefore need a monitoring BMS to keep them balanced. Usually the BMS will have other safety features, such as preventing overdischarging, short-circuit protection, or overcurrent protection. The nominal voltage is 3.3 V per cell, the full voltage is 3.75 V, and the empty voltage is around 3 to 2.5 V depending on manufacturer.

Using LiFePO$_4$ batteries will make an EV fast and desirable. The prices are now fairly reasonable, almost the same as nickel cells per watthour. LiFePo$_4$ batteries are the batteries most marketed for electric bicycle applications. The longevity of LiFePO$_4$ packs is still undecided because many manufacturers have not been around long enough. There is intense competition among rival manufacturers, and claims of thousands of cycles are everywhere but as yet unproven by significant numbers of real customers.

### 4.3.5  Future Power Sources: Fuel Cells and Flow Batteries

There are hundreds of different battery chemistries, many of which are only now receiving proper research and development attention because of the current interest in the environment, global security, and the price of fuel. The battery chemistries examined so far are just the main commercial battery chemistries that are used today, but there are also several less-well-known batteries used by experts for specialist applications. Fuel cells and flow batteries can be considered to be special cases of a battery in which the electrode plates are separate from the reaction medium, which is delivered to the cell from a separate store. This means that the reactants can be delivered continuously to the cell and not be limited by the size of the cell but stored in external tanks. Recharging these cells can be achieved simply by replacing or refueling the depleted tanks with fresh reactants. Air batteries, where one of the electrodes is air, are a type of flow battery.

Fuel cells are also a special case of a flow battery where one reactant is oxygen from the air that is used to react with a fuel (usually hydrogen). It is a common myth that fuel cells run on water; this is not true. Water is the waste product, and hydrogen is the fuel. While you can make hydrogen from water, it takes a lot of electricity to do so and is expensive.

Electric bicycles can be run on fuel cells just as easily as they can on batteries. New advances in fuel cells might be capitalized on by electric bicycles soon too. The early prototype fuel cells are all tested on electric bicycle applications first because electric bicycles are becoming a major market for manufacturers, and it's much cheaper to prototype with smaller vehicles than with cars. However, there are still several major obstacles to overcome with fuel cells.

Fuel cells are less efficient overall. Fuel cells are currently only 45 percent efficient, and hydrogen generation by electrolysis and compression is only 55 percent efficient (Shinnar, 2003). Both these steps are necessary to make hydrogen fuel cells work on vehicles, and the combined efficiency (25 percent) is no better than that of today's modern gas engines.

Fuel cells require pure gases. Pure fuels are expensive. Many fuel cells specify that 99.99 percent pure hydrogen must be used. If this purity of hydrogen is needed, then the efficiency of vehicles running on fuel cells likely would degrade quickly when they run on polluted city air. This would require the use of heavy cylinders of pure compressed air, which defeats the objective of using a fuel cell.

Hydrogen is very difficult to store, much harder than electricity. A complete fuel cell system including the hydrogen store usually will be beaten on energy density by most lithium battery chemistries. Metal hydride stores are too heavy. Compressed gases are too bulky.

Hydrogen is more difficult to make than electricity and more expensive.

Compressed gases are dangerous, especially hydrogen. You need a special license to carry cylinders in vehicles. No license is needed for batteries.

There is no distribution system for hydrogen. It would be unsafe and inefficient to convert natural gas pipelines to hydrogen pipelines (Shinnar, 2003). Currently, most experimental hydrogen filling stations use compressed cylinders delivered by truck, which is very carbon-intensive.

If renewable energy is stored as hydrogen instead of used as electricity, then we would have to build double the renewable-energy generating capacity (Shinnar, 2003).

### 4.3.6 Reviews of Batteries

Buying a battery over the Internet is all about trust and reputation. When you hand over the cash, do you trust the retailer not to rip you off and send you a broken battery? The battery has to perform for a long service life before you have gotten your money's worth. If you know a local bicycle shop that sells good batteries at a reasonable price, then you are lucky. Often you will be better off buying from a retailer and/or manufacturer with a good reputation than to get a cheap battery.

#### Ping (eBay Duct Tape LiFePO$_4$ Battery)

Ping is the only eBay retailer of LiFePO$_4$ batteries that has a solid reputation (as of 2010). There are two versions of the Ping battery, V2.5 and V3. Versions 1 and 2 have been superseded by the V2.5, but they are very similar; version 3 is a completely different battery. The V2.5 is a high-energy-density, low-power battery capable of only 1C continuous discharge and 2C peak. The V3 is a high-power battery. The BMS is good and protects the cells from everything you would expect and doesn't trip without reason. Be careful to select the size of battery that will fit the space on your bicycle. It's difficult to split your battery into smaller units because of the BMS circuitry. The batteries come without a case, with just duct tape and card to hold it together, so you must find a case yourself (Figure 4.27). Each battery on the Web site comes with a motor power suggestion that you should follow; otherwise, you will overstress the battery. These batteries are sold specifically for electric bicycles and scooters, so there is no hassle of making a battery out of cells or smaller packs. Ping includes a charger with every pack, which seems to be of good quality. The charger, battery, and BMS all work well together. It's a good whole-pack solution for novices or those who don't want a hassle. Ping's prices have increased now that the company has established a reputation, but the company's offerings are still one of the best deals on the market.

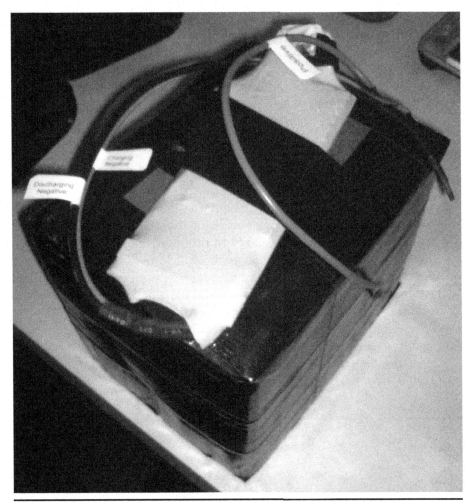

**FIGURE 4.27**   A brand new Ping battery. It just needs plugs for connecting.

### Zippy Flightmax Batteries (RC Lipo)

These batteries are made of Lipo cells and thus are lightweight high-power batteries, but they are not as safe as other battery chemistries. They are designed for RC airplanes, where power-to-weight ratio is critical. They have a very high discharge rate, so you don't need many to make a very fast bicycle. However, they require a special balancing charger that is not provided. No BMS is provided, so you have to find one yourself. The packs are very small. The largest pack size is 18 V, 4 Ah. Thus, for a decent electric bicycle, you would need about five of them. For an electric bicycle, you have to build a larger pack out of these smaller units.

## A123 Cells (LiFePO$_4$)

There are two sizes of A123 cell, 1.1 and 2.3 Ah. The 2.3-Ah cells are best for electric bicycles because you need fewer of them. It used to be difficult to acquire A123 cells, and people had to recycle power tool packs to get them. Now they are available secondhand on eBay. Some RC sites (Hobby City) also stock them new. It is harder to build a pack from these because they are often tabless, so welding is required. Some vendors are now offering prewelded custom packs. You have to connect up a third-party BMS and find a suitable charger.

## RC Battery Pack (NiMh)

The best deals locally on NiMh batteries are for RC hobbyists because they go through a lot of packs by hammering them on the race track. Their packs are smaller, and they can afford to buy new ones every few months. RC NiMh packs are usually discharge rated at 3C or 5C. These packs are harder to build into larger packs than the small lithium packs because you can't charge them in parallel. Because of this, it's quite difficult to build a decent-sized pack without using high voltages or separating the parallel strings for charging. You also have to buy an NiMh charger and thermistor, but a BMS is not needed. The price per watthour of these packs is also not that much cheaper than that of Ping packs. Table 4.1 provides a summary of commercial battery performance and value.

**TABLE 4.1**   Commercial Battery Performance and Value

| Brand Name | Battery Chemistry | Energy Storage, Wh/kg (Wh/lb) | Energy-to-Weight Ratio, Wh/L (Wh/qt) | Power-to-Weight Ratio (Continuous), W/kg (W/lb) | Capacity per Cost, April 2009, Wh/£ (Wh/$) | Format |
|---|---|---|---|---|---|---|
| Ping | LiFePO$_4$ | 100 (45) | 210 (210) | 100 (45) | 1.8 (1.2) | Pouch cells |
| Zippy Flightmax | LiPoly | 138 (63) | 285 (285) | 3000 (1364) | 1.4 (0.93) | Pouch cells |
| A123 | LiFePO$_4$ | 150 (68) | 180 (180) | 3000 (1364) | 1.04 (0.69) | Cylindrical cells |
| Powerizer | NiMh | 60 (30) | 300 (300) | 100 (45) | 2 (1.33) | Cylindrical cells |
| eBay SLA | Lead acid | 35 (16) | 90 (90) | 300 (136) | 4 (2.66) | Sealed plastic brick |

### 4.3.7 Resources for Buying Electric Bicycle Batteries

For up-to-date battery comparison tables, see www.wiki.saymoo.org/ index.py/EvdlGems/Batteries/LithiumIons.

For general battery use information, see www.batteryuniversity.com/ or www.buchmann.ca/.

### Electric Bicycle and Kit Retailers

All the electric bicycle retailers and kit distributors listed in Sections 2.1.4 and 3.4.1 will have their own electric bicycle battery suppliers. E-bikekit, eBikes.ca, and BatterySpace are good places to obtain batteries. Here you can also get materials for building battery packs, such as battery tabs, heat shrink, and BMSs.

- www.e-bikekit.com
- www.ebikes.ca
- www.batteryspace.com

To buy Ping batteries direct instead of from eBay, see www.pingbattery .com.

For A123 cells, Zippy Flightmax, and Turnigy Lipo packs and chargers, see www.hobbycity.com/hobbycity/store/uh_power_Search.asp.

### eBay Batteries

Lots of different electric bicycle batteries are available on eBay, including Ping, Highpower, and Cammy_CC. Or you can buy individual cells and build your own packs with Zippy Flightmax and A123 (used cells).

### Tool Packs

Cordless power tool packs are a great source of electric bicycle batteries. They often can be used without any modification needed.

- Bosch BAT836 fatpacks contain Konion cells.
- Dewalt lithium tool packs contain A123 cells.
- Makita power tool packs contain Sony 18650v cells.
- Milwaukee power tool packs contain LiMn cells.

## 4.4  Chargers and Battery Charging

One of the main problems people think of when they think of EVs is charging. The practical reality for most people is that charging is not a problem.

There are connections to the electricity grid everywhere, and all it takes is a few seconds to plug in. Charge times are not really an issue for most commuters because there is plenty of time when the vehicle is not being used and it could easily be undergoing charging. Is charge time a critical factor when buying a mobile phone? However, there are some people who desire much shorter charge times than the usual 1 to 4 hours. Charge rate of an electric car is limited by the plug voltage and the maximum amperage of the wall socket, which is usually 13 A. In countries with a 240-V supply, a 20-kWh electric car battery can be fully charged from empty in 6½ hours. With a 110-V supply, it would take twice as long as this from a standard socket. The charge time can be reduced if a dedicated circuit is connected to the circuit breaker by an electrician. This requires rewiring the house. As with an electric stove, a dedicated circuit will allow currents up to 15 A. To charge an electric car even faster would require higher voltage. Industrial three-phase power supplies are 480 V and 25 to 30 A, and these could charge a 20-kWh electric car in just over an hour.

However, because electric bicycles have much smaller batteries, they are not usually limited by the power of the wall socket but by the rating of the batteries. In theory, a 240-V, 13-A socket (3 kW) could charge your average electric bicycle battery in 10 minutes if the battery could handle the high charge rate. In countries with 110-V supply, it would take 20 minutes, twice as long. However, most batteries can't, in the long term, handle charge rates of more than 1C. This is why we still have the 4-hour charge times and 1-hour "fast charges." Even with the many promised advances in technology, this doesn't look like it is going to change anytime soon. The best way to improve charging therefore seems to be to have a bigger battery that can "mop up" charge quicker and approach that magic number wall socket limitation of 3 kW, which in electric bicycle terms is about 150 miles of range gained every hour. Charging at 150 mph sounds good to me!

## Battery Swapping

The other benefit of electric bicycles is that you can swap the batteries a lot easier than you can on a car. The company Project Better Place wants to see battery swapping stations, like gas stations, but where electric cars will swap their empty batteries for full ones in less time than it takes to fill a car with gas. There are many problems with this idea, but it does have the advantage of alleviating the limited range of EVs. The problem is that batteries are very expensive, and you would have to trust that someone is not giving you a dud. Some people have proposed battery rental as a solution, but then a company has to trust you not to break its batteries. This trust is never going to happen with DIY electric car owners and will require large automotive manufacturers to get onboard and standardize everything. Battery swapping is much easier with electric bicycles and can be done easily today when using bicycles with the same battery connections. Electric bicycles will lead the way in terms of battery swapping and rental and develop the principles and infrastructure that will transfer to future electric cars.

### 4.4.1 Charge-Termination Methods

Battery chargers can vary in sophistication in terms of how they detect the level of charge inside the cell or battery. The important part is how the charger determines when the battery is full and stops the charge. The first battery chargers were slow and unsophisticated, but now that electronics have progressed and have become cheaper, much more accurate methods of charge termination are used. There are two basic types of charging, constant voltage and constant current. Constant-current charging is used at the start of charging when the voltage difference between the cell and the charger would be too high and would lead to excessive currents. Control of the charge current in this way controls how fast the battery is charged and is necessary for fast charging. Constant-voltage charging is used to accurately top up the battery to ensure that no damage is caused by overcharging. Constant-voltage charging is usually slow and occurs at the end of the charge profile, once 80 percent of the charge has been delivered. As the battery reaches the same voltage as the charger, the rate of charge drops to zero.

The most primitive form of battery charging is trickle charging. Trickle charging involves a very low constant current with no endpoint determination. The user has to judge when the batteries are full by timing the charge and then manually disconnecting the batteries. This is a poor method of charging because it requires the user to know about the state of charge of the battery and calculate how long is required to charge it and then to remember when to turn off the charge. Because of these pitfalls, trickle chargers are designed to very slowly charge the battery to prevent destructive failure if the charger is left on too long. Most battery chemistries are damaged by continued trickle charging past their full state of charge. The only exception to this is the NiCd battery, which is why it is used in backup power applications such as emergency lighting, where it can be trickle-charged to prevent self-discharge. NiMh cells also can withstand some trickle charging. Some electric bicycle battery chargers have a short trickle-charge phase after the main fast-charge period to try to get a bit more capacity into the cells.

Nickel batteries have a special charge-determination method based on temperature. Modern electric bicycle NiMh chargers use a thermistor located in the battery pack that determines when the pack is fully charged. This determination will be be based on either rate of temperature increase or a set temperature limit. If the battery is discharged too quickly during a ride, the temperature of the battery will rise, and the charger may refuse to charge the battery until it has cooled down. In theory, NiCd and NiMh chargers are almost interchangeable. However, the trickle charge on some NiCd chargers is too much for NiMh cells, so old NiCd chargers should not be used for NiMh cells, but it's okay to use NiCd cells in modern NiMh chargers.

### 4.4.2 Inside a Battery Charger

A battery charger is basically just a means of converting electricity from one form into another so that it can be used to charge a battery. A battery charger is very similar to a switched power supply, such as a laptop power supply. The difference is that a battery charger has a method of detecting full charge and switching off. Switched power supplies are an advance over the old big transformer power supplies because they use fast switching to reduce the size requirements of the transformer needed in the power supply. This advance in electronics has made the power supplies cheaper and smaller, which has made it possible to have cheaper consumer electronics and computers in every home.

The switched power supply works as follows: First, the main circuit alternating current (ac) electricity is converted to direct current (dc) by a bridge rectifier; then the dc is switch quickly on and off using a special transistor. The fast switching is much faster than the 60-Hz frequency that main circuit transformers use. Since induction is proportional to the rate of change of current, this fast switching allows the transformer to be much smaller than a main circuit transformer for the same power rating. The transformer changes the voltage to the desired voltage, and then big filter capacitors are used to smooth out the peaks caused by the switching. The interesting thing about switched power supplies is that they work with dc or ac power because it is all turned into dc anyway right at the start. Therefore, you can do interesting things such as charge your electric bicycle from a bank of solar-powered batteries using your electric bicycle charger, for example (see Section 6.3.2).

There are other possible power supplies that can be used for charging batteries. A *buck converter* is another power supply that uses switching but for a different purpose. A buck converter has the advantage that is doesn't contain a transformer at all, which saves on weight and cost (hence the name). A buck converter consists of a method of switching, usually a special transistor such as a MOSFET, coupled with a method of smoothing, such as a capacitor. In a buck converter, the ratio of on to off in the switching determines the voltage that will be supplied. It's possible to change the on/off ratio of the switching to get any voltage below the main circuit supply voltage. It's possible on transformer-based power supplies to have variable voltage too, and this is usually accomplished by taps in the transformer winding.

The advantage of the buck converter is that it's lightweight and can be carried round on an EV as an onboard charger more easily than some other chargers. The disadvantage of the buck converter is that it is not isolated from the main circuit power like with a transformer-based power supply. Isolation of the power supply is key to preventing and reducing the severity of electric shocks if they occur. Some electric car owners have made their own "bad boy" chargers based on buck convertors because they want to avoid the high cost of commercial electric car chargers. Electric cars are still a niche market all over the world, and therefore, specialized equipment costs are still quite high. Electric bicycles, by comparison, are not a niche market; millions of units are sold annually in China and other developing countries, so the costs are much lower and there is no need to build DIY chargers.

# CHAPTER 5

# How to Maintain and Repair Electric Bicycles

## 5.1 Equipment for Maintaining Electric Bicycles

### 5.1.1 How to Use a Multimeter

The multimeter is the most useful diagnostic tool for electric bicycle maintenance, and it's essential to learn how to use one and have some basic understanding of electricity. A multimeter can be used to measure all the important properties of circuits, including voltage, current, and resistance.

Some fancy multimeters can do other things too, such as measure temperature, frequency, etc. The main dial is usually the only input that is required by the user, so it's really a very simple tool. The dial is used to change the range and function settings of the meter. Each measurement is split into different ranges that the meter can detect. For voltage, the dial will be split into millivolts (200 mV), volts (2 V), tens of volts (20 V), and hundreds of volts (200 V). The numbers around the range-select dial (and here in parentheses) correspond to the maximum measurable value on that setting. The setting you use also gives you the multiple of the reading you see. In other words, a voltage reading of 19 on the 200-mV setting will mean 19 mV; a resistance reading of 24 on the 200-M$\Omega$ setting will mean 24 M$\Omega$.

Don't be afraid to have a go and measure all kinds of stuff. You can't damage the meter by having it on the wrong range or the wrong setting, except using the wrong probe plug-in holes. Obviously, don't mess with high-voltage electric, though. If you don't know what the reading will be, select a higher setting than needed, and this will give you a rough "ballpark reading." Then select the appropriate lower setting to get a more accurate reading.

Resistance, direct-current (dc) voltage, and alternating-current (ac) voltage all use the same plug-in holes, COM for the black lead and V for the red lead. With these plugged in, the meter has a very high resistance and will not cause any problems for the stuff you are measuring. To measure current, however, you need to change the lead plugs over to AMP for the red lead and keep COM for the black lead. Here, the meter will have a very low resistance. A common mistake when measuring battery current and voltage is to forget that you have the leads plugged for current measurement and attempt a voltage measurement. This causes the meter to act as a short across the battery and will vaporize the contacts on the meter and the thing with which you complete the circuit. This can cause damage to plugs, etc.

In an ideal multimeter, the resistance would be infinitely high when on voltage measurement in parallel with the object to be measured and zero when on current measurement in series with the object to be measured. However, real meters do not have infinitely high resistance, and this sometimes can cause confusion when measuring volts dropped across very high-resistance components close to the limits of the meter. This is so because the current may flow through the multimeter, thus altering the circuit being measured.

### Multimeter Continuity Testing

A useful function of a multimeter is continuity testing. *Continuity testing* is where you use the resistance measurement function to check what's connected to stuff. If you have a bundle of wires coming out of a cable and you want to find the one that's connected to your lights, you can use the continuity test. Just put a probe on the wire that you think ends at your lights, and then test each of the wires coming out of the cable at the other end. This is where the crocodile lead attachment is useful. When the meter shows zero resistance (or close to it), you have found the wire that's connected (Figure 5.1); a one means that the measurement is off scale or infinite resistance (Figure 5.2). Some more expensive meters have a useful audible bleep function that does the same thing, but you don't need to look at the meter while checking. Sometimes with cheap meters there is an offset reading, so it may read 0.5 V even when the meter probes are firmly contacting each other. Take this into account when testing. For accurate readings of low resistance, you need more specialized tools such as an equivalent series resistance (ESR) meter.

### Multimeter Polarity Detection

Another useful multimeter feature is polarity indication. This allows you to find out which way is positive and which way is negative on unmarked batteries or chargers and saves you from blowing stuff up by connecting things the wrong way. Make sure that you have the red lead in the red V port and the black lead in the black COM port. Then measure the voltage of a battery or other circuit. If you connect the leads and see a positive voltage, then the

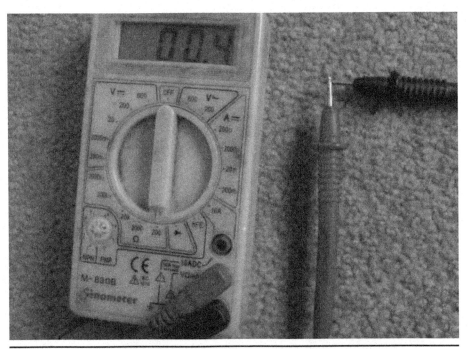

**FIGURE 5.1** Closed circuit shows as 0 (0.4Ω is close enough).

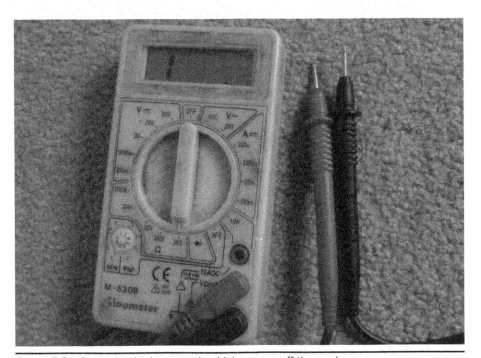

**FIGURE 5.2** Open circuit shows as 1, which means off the scale.

red probe is touching positive and the black probe is touching negative (Figure 5.3). If you see a negative reading, then the opposite is true (Figure 5.4). If you get confused with your polarity test, then test a clearly marked household battery for comparison. Also, swapping the probes around a few times

**FIGURE 5.3** A positive reading shows that whatever the red probe is touching is positive.

**FIGURE 5.4** A negative reading shows that whatever the red probe is touching is *not* positive.

will help. Never connect stuff if you don't know the correct polarity. Don't take guesses based on the color codes of wires because they might be from different manufacturers and mean different things.

You can't measure the voltage of the meter's power source using the same meter. It uses the battery voltage to compare against the circuit voltage and gets confused. If you suspect a dodgy reading on the multimeter, change the battery and try again.

You can't usually measure currents higher than 10 A with a multimeter, which is insufficient for electric bicycle measurements. To measure currents higher than this, you need to use a shunt resistor. A shunt resistor or clamp meter is a known-low-resistance element with good heat dissipation. The current is run through the shunt resistor, and you measure the voltage drop across it. From this voltage reading, the current is calculated, which usually is done by a panel meter to give a convenient reading. You don't have to buy a shunt resistor. The shunt can be anything, even a section of wire in your power leads. All you need is to know the resistance to be able to use Ohm's law, the relationship between voltage, current and resistance. You can calculate the resistance again by using Ohm's law and measuring volts and amperes either at a current below 10 A or by using a clamp meter or power meter to compare.

**Disclaimer:** Don't stick multimeters in main circuit power sockets.

### 5.1.2 Power Meters

Power meters come in different forms. Battery power meters can be used onboard an electric bicycle to report the power consumption in real time to the operator. Socket power meters can be plugged into a socket to measure the power consumption at the socket being used to charge the vehicle. A power meter is a more specialized tool than a multimeter and is solely for the purpose of measuring power consumption. Most battery power meters are marketed at remote control (RC) hobbyist sites (Figure 5.5), for example, the Watts Up Meter or Doctor Watson (both available at www.reuk.co.uk/Watts-Up-Meter.htm).

There is now a power meter designed specifically for electric bicycle use—the Cycle Analyst. The Cycle Analyst is best because it usually can plug straight into your controller and use its shunt, which means that you don't have to run heavy-gauge power wires up and down the handlebars. The RC power meters have the shunt built into the meter, so you have to run a power wire through the meter. Although it is possible to hack apart the cheap meter and relocate the shunt to a more convenient location. The Cycle Analyst also can measure the speed you are going and calculate the electric consumption per mile. Battery power meters will show you instantaneous watts, volts, and amperes, as well as count the watthours and ampere-hours used and the ampere peak and voltage minimum. This information is es-

**FIGURE 5.5**    A very useful and easy to use RC-style power meter.

sential for checking that there are no problems and for finding how much battery capacity you have left.

A socket power meter will tell you volts, amperes, instantaneous power, energy used, and power factor. Most can be set up to tell you how much money the power is costing. Socket meters are cheap, but sometimes the units are kilowatthours, which are too big for accurate determination of electric bicycle power use. Power meters always should be connected between the battery and controller, not after the controller.

### 5.1.3  Panel Meters

Panel meters (Figure 5.6) are so called because they are designed to mount on an instrument panel. You can get cheap panel meters to measure all sorts of different things—volts, amperes, temperature, frequency, and count—but it's not really possible to get more advanced measurements such as speed, power, watthours, etc., so they are not so good for electric bicycle meters. The main advantage of panel meters is that they are cheap; for all other purposes, it's best to get a power meter.

Most panel meters can measure only one property, but some are dual-function meters with a button that toggles between functions. Some panel meters are blank liquid-crystal display (LCD) panels, which can be configured in many ways and need to have resistors added to make them measure the property you want. This is very complicated and best avoided. The eBay panel meters advertised as voltmeters, ampmeters, etc., are simple to use and already preconfigured.

**Figure 5.6**    Three panel meters. These meters are cheap, but they are not as easy to use or as useful as RC-style power meters.

There are both LCD and light-emitting diode (LED) panel meters. The LED meters look more low tech but can be better for reading from angles; LCD meters look a bit more professional. You can change the measurement range of some panel meters by soldering different resistor values on the back, but it's best to buy a meter with the right range in the first place. Usually panel meters have bare-bones circuitry on the back, so they have to be mounted into some kind of box for electric bicycle use. This makes them bulkier than battery power meters.

To measure current with a panel meter, you have to have a shunt, and the panel meter basically measures the voltage drop across it just as described earlier for the multimeter. Thus current panel meters will input a millivolt signal and display a current proportional to that signal. You have to buy the right resistance shunt; otherwise, the millivolt per current signal given to the meter will be different, and the meter will be off by some degree.

### 5.1.4 Clamp Meters

A clamp meter (Figure 5.7) is another way of measuring current. Clamp meters use Hall sensors to measure the magnetic field generated by the flow of current. You put the wire through a hole in the meter, and the Hall sensor measures the current. You need to put a single wire through the metering

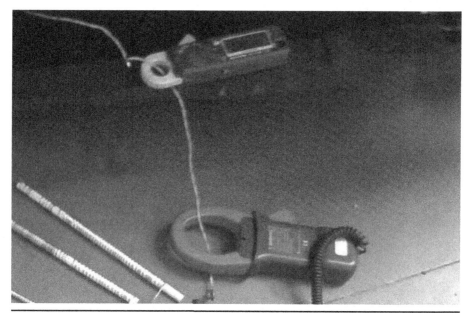

**FIGURE 5.7**   Two different styles of clamp meter.

hole; you cannot put both positive and negative wires through because the fields will cancel each other out. This causes a problem with insulated cables because you can't simply clamp the whole cable. You have to expose a single wire, which defeats the point a little.

A great application of clamp meters that I recommend you buy is the wireless home energy monitor (Figure 5.8). These monitors provide a portable display of your home energy use, which is wirelessly connected to a clamp meter that you put around your house electricity meter box. The display lets you know instantly how much power your house is using. You can use it to check how much power your separate appliances are using just by switching them on and off and looking at the numbers change. Most of the energy monitors also have a data-logging function that adds up all your energy use and calculates how much your electricity bill will be. This is great for people who are interested in saving costs or lowering their carbon footprint. In addition, an energy monitor allows you to demonstrate to your housemates/girlfriend/wife/parents how little electricity your electric bicycle really uses in comparison with such things as your refrigerator, which uses 10 times the energy of your bicycle. The monitor has other advantages, too, such as only one check to know that all your lights and appliances are off before you leave the house. You can use an energy monitor to measure dc electricity as well, but you need to buy a separate adapter.

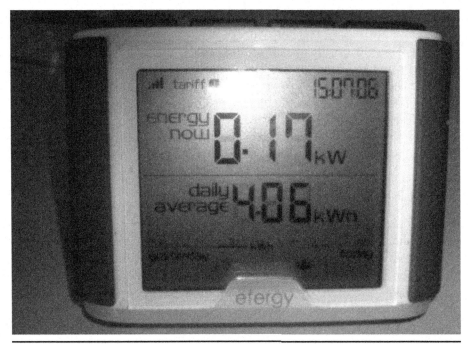

**FIGURE 5.8**    Wireless home energy monitor. This monitor communicates wirelessly with a clamp meter on the supply cable in the meter box.

### 5.1.5  Bicycle Speedometers

Bicycle speedometers work by counting the frequency of a signal generated by a magnet spinning around the wheel through the path of a sensor. The dimensions of the wheel have to be accurate; otherwise, the displayed speed will be wrong. It's amazing how many people don't do this correctly, and the end result is kids telling you that they can do 40 mph on their BMX bicycle because the speedometer is set for 26-in wheels and they have 20-in wheels.

### 5.1.6  Thermistors and Thermocouples

These are two different types of temperature-measuring probes that can be useful on electric bicycle projects. If you are pushing the limits of a motor or a battery, then you will want to monitor its temperature, at least initially, to be sure that you are not operating outside its safe working limits. Thermistors have been mentioned before and are resistors that change resistance based on temperature. You can buy cheap aquarium thermometers that contain thermistors for monitoring temperature usually up to 70°C. You also can just use a multimeter and draw up a calibration plot of resistance against temperature. The resistance drops as temperature increases.

Thermocouples are different in that they generate a millivolt signal depending on temperature. Thermocouples are better because they have a much wider range of operating temperatures, something that I hope you won't be need for electric bicycle use because if your motor gets that hot, it will have toasted itself! To measure thermocouples, you need special meters that are bulky for electric bicycle use. It is best to stick to thermistors and aquarium thermometers—they are easier.

### 5.1.7 Data Loggers

Data loggers (Figure 5.9) can be useful for battery testing. They give more information than a simple power meter. Data loggers can be used to log the volts from a battery for measurement of capacity during charge or discharge. When the battery voltage drops away, it is empty, and if you use a constant-current load, then you can work out the capacity of the battery. However, do not let your battery voltage drop below the minimum safe limit because it will damage the battery. You have to monitor the battery closely or fit a low-voltage cutoff circuit to it. You can use light bulbs for a constant-current load.

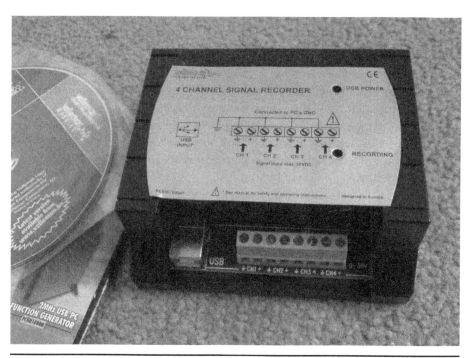

**FIGURE 5.9**  A USB data logger can be useful for very advanced projects. Remember that the negative terminals are all connected together.

Most data loggers have at least four channels, so you can improve the accuracy by logging current too. Just use a shunt, and measure the voltage across that. Then do the math in Excel. Be careful, though, because most data loggers have their channels share a common negative, which means that it's easy to short circuit stuff if you don't think the connections through. A mistake could cause damage to the computer with which you are using the data logger. Use a plug-in clamp meter for the current measurement if you are not sure. You can, if you want, make the system mobile by using a laptop and data logging your whole ride. This is only useful for academics, though.

### Data-Logger Tips

If the data logger accepts readings only in a specific low-voltage range, then you can use a potential divider circuit to reduce the battery voltage down to the level needed for the data-logger signal. Just put the battery voltage across two series resistors, and connect the data logger across one of them. Done with two equal resistors, this will halve the signal to the data logger. Use Ohm's law (see Section 5.1.1) to work out the resistor values that are needed.

Be careful what you measure. If the data logger is connected to a computer that is plugged into a wall socket and you measure something that's also plugged into a wall socket, then you sometimes can cause a ground leak that throws the circuit breakers in the house. It's best to use a battery-powered laptop for this purpose.

## 5.2  Diagnosis: Finding the Problem

### Things that go wrong on electric bicycles.

The most common things to go wrong on electric bicycles are the mechanical parts. Mechanical problems usually are obvious, and you can get those solved at your local bicycle shop, so I won't address those issues here. The second most likely things to go wrong are the cheap plastic connectors and wiring used in most of the kits. After this, the electronic parts—controllers, chargers, and battery management systems (BMSs)—sometimes may go wrong, but always start the fault-checking process by suspecting the wiring. Do the simple checks first before you check the complicated or expensive things; otherwise, it's possible to mess things up during the process of fault checking. Usually, if a problem has presented itself before, then it will be obvious, and you will know instantly what to do to correct it. If you don't know from the symptoms, then process of elimination is the only way to find a problem. Table 5.1 is a quick reference list of symptoms that you can run through to narrow down the problem to a few possibilities. Table 5.2 is an ordered list of detailed tests that by process of elimination should show you exactly what's wrong with your electric bicycle. You will find out how to fix these problems in Chapter 6.

**TABLE 5.1**  Diagnosing Problems from the Symptoms: A Quick Run Through This Table Will Narrow Down the Problem to a Few Possibilities

| Symptoms List | Possible Causes |
| --- | --- |
| Won't move; nothing works; no voltage on power meter; no light on controller | Fuse blown, melted fuse holder, bad connection along the battery power line, tripped the BMS cutoff, dead-cell open circuit |
| Won't move; voltage on meter; controller light on | Blown capacitors, blown field-effect transistors (FETs), connections to motor unplugged |
| Starts moving but cuts out as soon as you turn the throttle | Broken BMS tripping, very weak or dead cell in the battery, resistance between cells dropping cell voltage and causing BMS to trip, balance lead has broken off inside battery |
| Moves but not smooth; loss of power and very jerky | Broken Hall sensor connection, broken motor phase connection, incorrect Hall/phase connection |
| Lack of power or range | Charger not working and battery is flat, weak cell or cells in battery |
| Runs well but cuts out after so long | Melted fuse holder, overheating controller thermal protection, overheating batteries' thermal fuse, bad low-voltage charger |
| Runs but makes a strange grinding noise coming from the motor | Motor bearings broken, motor gears broken, debris in the motor |
| There is smoke coming from the batteries! | Short circuit inside the battery pack, damaged battery insulation, broken charger overcharging the battery |

If your bicycle doesn't work, go through Table 5.2 to find the problem. It is set up with the simplest and most important tests first.

### 5.2.1 How to Check Connections and Connectors

The connections in some Chinese electric bicycle kits are very cheap and prone to failure. They can look okay but be broken inside. To check the connections, it is sometimes possible to backprobe the pins in a connector. Put

**TABLE 5.2** How to Test Each Component Using a Process of Elimination to Find the Cause of the Problem

Start from the top and work your way down until you find the problem. All these tests require the use of a multimeter, so read Section 5.1.1 if you are unsure of how to use one. A more detailed explanation of how to do the tests and what the results mean is included next.

| Item to Test | How to Test It | A Bad Test Result Is… | Possible Cause |
|---|---|---|---|
| Battery | Test battery voltage with multimeter between the positive and negative wires | 0 V | Fuse blown, melted fuse holder, bad connection along the battery power line, tripped the BMS cutoff, dead-cell open circuit |
| | | Lower voltage than normal | Dead cells, flat battery, weak cells, balance lead has broken off inside battery |
| Motor | Test motor coil resistance with multimeter between each phase wire | Open circuit between any two phase wires (multimeter shows 1) | Broken motor windings, power leads snapped inside motor |
| Controller | Test resistance between each phase wire (motor side) to each power wire (battery side) | Short circuit (zero resistance) | FETs are blown inside controller and need replacing |
| Power lead | Inspect and continuity check power wires from battery to controller | If anything looks melted, corroded, or doesn't fit right or the resistance is more than 0.5Ω | Bad external fuse or holder, corrosion damage to connectors |
| Hall sensor lead | Inspect and continuity check each Hall sensor wire from motor to controller | If anything looks melted, corroded, or doesn't fit right or the resistance is more than 0.5Ω | Bad wire or plug needs replacing |
| Throttle lead | Inspect and continuity check throttle wires from throttle to controller | If anything is corroded, broken, or doesn't fit right or the resistance is more than 0.5Ω | Bad wire or plug needs replacing |

*(continued on next page)*

137

**TABLE 5.2** How to Test Each Component Using a Process of Elimination to Find the Cause of the Problem (continued)

| Item to Test | How to Test It | A Bad Test Result Is... | Possible Cause |
|---|---|---|---|
| Three phase wires | Inspect and continuity check each phase wire connection from controller to motor | If anything looks melted, corroded, or doesn't fit right or the resistance is more than 0.5Ω | Bad wire or plug needs replacing |
| Controller Hall sensor power | Check voltage between red and black Hall sensor wires from the controller when switched on | There should be 6 to 12 V across the red and black Hall sensor wires from the controller; if not, there is a problem | Broken controller |
| Motor Hall sensor switching | Check voltage between each of the blue, green, and yellow Hall sensor wires and the red Hall sensor wire when plugged into the controller and switched on while spinning the motor | Voltage should switch between high (6 V) and low (0 V) as the motor is spun; if not, there is a problem | Broken or disconnected Hall sensor |
| Fuse holder | Inspect and continuity check battery fuse and holder | If fuse holder looks melted or the fuse doesn't fit right or the resistance is more than 0.5Ω | Bad fuse/holder needs replacing |
| Battery charger | Check voltage of battery charger when plugged in | If lower or higher than the fully charged battery voltage | Broken charger; if voltage is low, then battery may be flat or unbalanced but otherwise okay; if voltage is too high, then the faulty charger may have over-charged the battery and caused damage |
| Cells | Check voltage of each cell | Cell is bad if 0 V or lower than the nominal cell voltage; pack is unbalanced if any cells are more than 0.2-V different | Dead or weak cell needs replacing, disconnected balance wire |
| BMS | Check cell voltages under load | BMS is bad if the cell voltages are okay under load but the BMS still cuts out | Faulty BMS needs replacing |

your multimeter probes in the back of each pin going through the connector and to the back of its mating pin on the other side of the plug. Measure the continuity of the connection. You may need to use a stinger—a piece of wire connected to the meter probes. If it is not possible to measure continuity in this way, then it is necessary to scrape away some of the insulation on either side of the connector to measure the continuity of the wires going through. You should test each wire because sometimes individual pins inside connectors can fail. Reinsulate the wires carefully when finished. Always give the connections a thorough visual inspection; it's usually possible to spot melted or corroded connections.

### 5.2.2  How to Check the Battery

If you check the voltage (Figure 5.10) of the battery and it is either 0 V or lower than normal or intermittent, then this indicates that there is a battery problem. The first thing to check is the fuse and fuse holder; hopefully, it is one of them. Check to see if the fuse or fuse holder is melted. A melted fuse obviously will result in an open circuit. A melted fuse holder can result in an intermittent connection or a high-resistance connection. During use, the melted fuse holder may work initially and show a voltage, but it can heat up and melt further causing the bicycle to cut out.

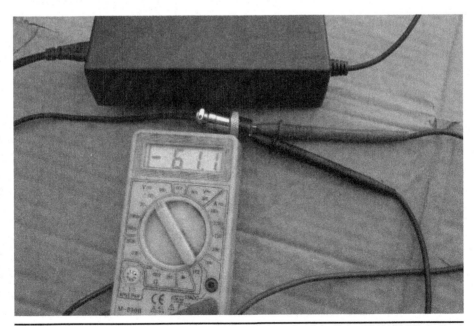

**FIGURE 5.10**  Checking that the battery charger voltage is okay. For a 48-V LiFePO$_4$ battery, you want to see around 61.5 V from the charger. Ignore the negative sign; this just means that the probes are attached the wrong way.

After ruling out the fuse, check that the charger is working. Connect a multimeter across the charger output pins, and it should show a voltage equal to the fully charged battery voltage. If the voltage is low, then it may not be charging the battery, and the battery may be flat or unbalanced but otherwise okay. This may present as a decreased range if the charger is low voltage or lack of power if the battery is flat and the charger is completely broken. If voltage is too high, then the faulty charger may have overcharged the battery and caused damage. If there is a BMS, it may have cut out to protect the pack.

If the fuse and battery charger are both okay, then next check the cell voltages to see if any cells are dead or any balance wires are disconnected. This is easy on a pack with a BMS. Just open up the insulation on the BMS, and check the cell group voltages from the balance wires (Figure 5.11). Measure from battery +Ve to each balance wire, and write down all the voltages. Then subtract the previous reading from the current reading, and the difference will be the voltage of that cell. This is the preferred method of measuring voltages. You also can just put the multimeter probes across neighboring balance wires, but usually they are too close together, and it's too easy to slip and cause a short circuit.

All the cells should be the same voltage. They should all be around 3.3 V if they are charged LiFePO$_4$ cells and 1.35 V if they are charged NiMh or

**FIGURE 5.11**    Check the BMS. If your battery is a lithium battery, it should have a BMS. All the cell voltages can be measured from the white multipin connector at the bottom.

NiCd cells. If any cell is below 2.5 V for LiFePO$_4$ or 0.9 V for NiMh/NiCd cells, then the cell is dead, and that is the problem. A dead or weak cell will cause the BMS to trip. If a balance wire is disconnected, it will show as 0 V on one of the cells. A disconnected balance wire will trip the BMS. To reset the BMS, you have to unplug all loads from the battery. If a cell is dead, then it will need to be replaced to stop the BMS from tripping. When you have finished, either heat shrink the BMS back up or duct-tape it closed with card insulation. For batteries without a BMS, you will have to break open the heat shrink on the battery to check the cell voltages. If you cut the heat shrink carefully, it can be duct-taped back together and look respectable.

If all the cells show the same voltage or are within 0.2 V of each other, then next check the BMS. Do this by repeating the same cell measurements but under load to see the voltage sag of the cells. If a cell sags a lot when a small load is applied, then it is weak and causing the BMS to trip. If the cell voltages are all okay and the BMS is still cutting out anyway, then by process of elimination the fault is with the BMS. Start with a small load (a light bulb), and then build up to a bigger load (lots of light bulbs or higher-wattage bulbs) and see when the BMS cuts out (Figures 5.12 and 5.13).

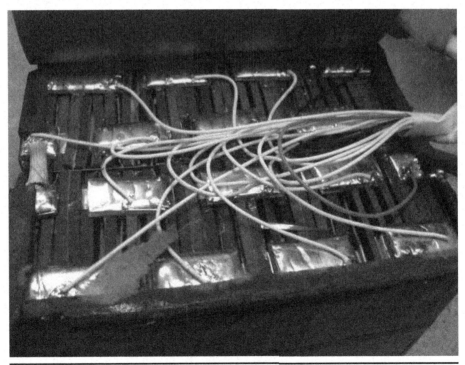

**FIGURE 5.12**    The cells are all connected to the BMS via the white balance wires. If one comes off, the BMS will cut out to protect the battery. The solder mass joins the cells three in parallel and then in series to the next cell group.

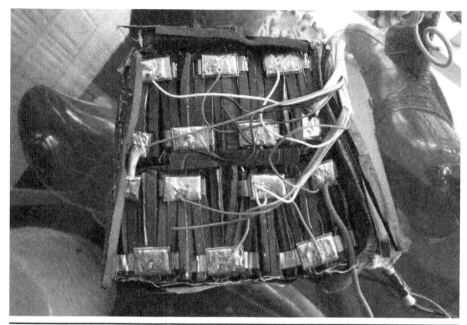

**FIGURE 5.13**    Look for obvious signs of damage. Notice that the ballooned shape of the cells distorts the pack.

### 5.2.3  How to Check the Motor

#### Check Winding

The first thing to check on the motor is the coil resistance (Figure 5.14). Use a multimeter set to resistance measurement, and connect it across the motor phase wires. With a brushed motor, there are only two wires to check, so connect the multimeter there with crocodile clips. Inside the hub, the windings are switched between many different sets as the wheel rotates. Rotate the wheel slowly to see if the resistance changes as the brushes switch over. If the resistance is 1, an open circuit, then there is a problem with the motor, and it should be opened for further inspection. In low-power motors, the winding resistance is higher and should be measurable by a multimeter. If the resistance of the coils drops during a small segment of the rotation, then there is probably a problem in which the coils have overheated and shorted together.

With a brushless motor, there are three wires, and you should check the resistance between each two wires, that is, three combinations. If any coil resistance shows 1, an open circuit, then there is a problem with the motor, and it should be opened for further inspection. These symptoms can result if the motor has been spun inside the dropouts that caused shearing of the wire. If the resistance is lower on one phase than it is on the other two phases, then this also indicates a motor problem. The motor may have over-

**Figure 5.14**  Checking that the windings are okay.

heated and melted insulation, causing a short circuit, or melted a phase, causing an open circuit. It can be difficult to measure the coil resistance accurately on a powerful motor because it will be very low and may be difficult to distinguish from a short circuit when using a cheap multimeter. An ESR meter might help to better identify motor coil problems.

### Check Hall Sensors

If the phase wires seem to be intact, then the next thing to check is the Hall sensors if the motor is brushless. The Hall sensors should switch between high (6 V) and low (0 V) as the motor is spun. The Hall sensors need to be supplied with power from the controller in order to do this, so you have to plug the controller into the battery and to the Hall sensor cable and switch it on. You don't need to plug the motor phase wires in. Use the multimeter on voltage setting to measure between each of the blue, green, and yellow Hall sensor wires and the red Hall sensor wire while spinning the motor by hand. Voltage should switch between high (6 V) and low (0 V) as the motor is spun. If any Hall sensor wire voltage does not switch, if it is stuck on 6 or 0 V, then that Hall sensor may be damaged and need replacing. Check that the controller is supplying the voltage to the Hall sensors by measuring the voltage between the black and red Hall sensor wires. It should be around 4 to 12 V; if it is 0 V, then there is a problem with the controller.

In a brushed motor, it's more difficult to check the brushes. The motor has to be disassembled and examined. If there is a problem with your brushed motor—for example, it has lost power and is running hot—test the windings as indicated earlier, and then follow the disassembly procedure mentioned in Section 5.4.6. Look for burn marks caused by sparks arcing or buildup of carbon dust from the brushes on the contactor plate. Disassemble the brushes, and check for damage. Make sure that they slide smoothly inside their housings.

### Check Hall Sensor Sequence and Timing

To check the sequence and orientation of the Hall sensors, there is another test you can perform. This one requires building a small LED board (Figure 5.15). The idea is to visualize the open/close of the Hall sensors with LEDs.

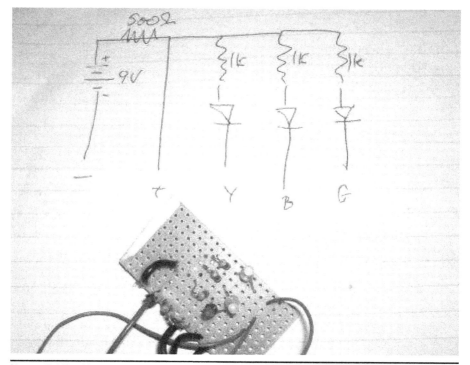

**FIGURE 5.15**    Circuit diagram of the Hall sensor checker and the finished item on a copper strip board.

If there is a time when all LEDs are either on all at once or off all at once, then the motor has 60-degree Halls. If the sequence shows only two LEDs on at any one time, then the motor has 120-degree Halls. You can use this to set the jumper settings in the controller for what type of motor you have. The LEDs should light up in a sequence one after the other when they are in the correct orientation. You can use this to find the correct orientation, and this will narrow down the 36 possibilities when trying to match a motor and controller wiring.

1.  Get a piece of bread board or copper strip board, three different colored LEDs, five 1-k$\Omega$ resistors, and a 9-V battery or power adapter.
2.  Next, connect the circuit as shown in Figure 5.15.
3.  Cut two lines through the copper strips.
4.  Position the resistors across one of the cut traces and the LEDs across the other cut trace in series with the resistors.
5.  At the positive end of the LEDs, the side without the flat part, tie all the strips together.
6.  At the negative side of the LEDs, after the resistors, add wires that lead to the Hall sensor wires. Add crocodile clips, or whatever, to attach to the Hall sensor wires.
7.  Tie the positive end of the power supply, through two parallel 1-k$\Omega$ resistors, to the LEDs.
8.  Attach crocodile clips to the negative of the power supply so that you can hook it to the black Hall sensor power lead.
9.  Attach a lead with a crocodile clip to the positive end where all the junctions meet. This will hook onto the positive Hall sensor lead.

### 5.2.4 How to Check the Throttle

There are two types of throttles used on electric bicycles, a potentiometer or a Hall throttle. Follow this procedure to test an unknown throttle type or to check for a damaged throttle:

1.  First, use a multimeter to test the throttle. Check the resistance across all the pins (three combinations). Check the resistance while you twist the throttle and look for changes.
2.  If there is a change in resistance when you twist the throttle, then this is a potentiometer-based throttle. If not, then this is either a Hall throttle (likely) or a broken throttle.
3.  Next, find an old mobile phone charger or batteries, around 4 V, and wire this up to the throttle, red throttle wire to positive and black throttle wire to negative. Put a 100-$\Omega$ resistor in the circuit on one of the

power supply leads. If you don't know the polarity of the phone charger, then use a multimeter to polarity check it (see Section 5.1.1).

4.   Set your multimeter to the 20-V setting, and measure between the green sense wire and the black negative wire or red positive wire and twist the throttle while you do this.

5.   If there is a voltage change, then this is a working Hall throttle. If you don't see any changes on either of these tests, then the throttle is dead.

### 5.2.5  How to Check the Controller

The two main faults that can happen with the controller are that the field-effect transistors (FETs) can short circuit or the capacitors can blow. To check that the capacitors in the controller are not short circuited, connect a multimeter in resistance mode across the controller's positive and negative battery connections. The resistance should be infinite or very high. To check that the capacitors have not been blown open circuit, connect the controller to the battery while a voltage meter is connected across the same controller battery power leads. There should be a small spark as the capacitors charge up, and the voltage should rise quickly to the battery voltage. Then disconnect the battery, and the voltage should drop slowly to 0 V. If the voltage drops instantly to 0 V, then it's likely the capacitors are blown and need replacing.

Next, test that the FETs in the controller have not been shorted. You need to test resistance between each phase wire (motor side) and each power wire (battery side). There are six combinations. The resistance should be high or infinite on all the combinations (Figure 5.16). If any show a resistance lower than 1,000 $\Omega$, then there is a short circuit, and the controller will need some work to fix it.

## 5.3  Maintaining Electric Bicycle Components

### 5.3.1  Maintenance of a Battery Charger

When plugging in an electric bicycle battery charger, there is usually a recommended sequence of doing things depending on the type of battery and charger, so read the manual if you have one. The open-circuit voltage of some battery chargers will "float" up much higher than the battery voltage if it is plugged into the main circuit first with no load before plugging it into the battery. This is true with many of the older car battery chargers. Then, when the battery is plugged into the charger, the voltage difference and low resistance cause a spark to jump across the connectors and partially melt them. The spark also can damage capacitors in the charger. It therefore used

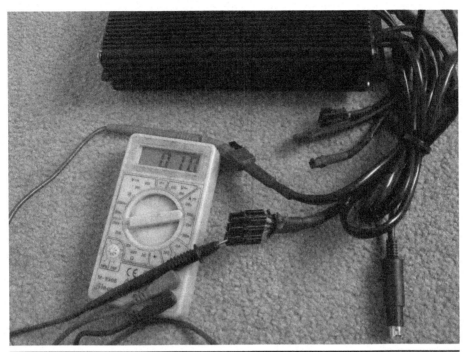

**FIGURE 5.16**    Checking the FETs. In this figure, the blue phase shows no sign of short circuit to the positive supply with 10 MΩ of resistance.

to be advised to plug into the battery first before plugging into the main circuit. However, most modern chargers will only float up to the end of charge voltage when plugged into the main circuit if not connected to a battery. EV packs are much higher voltage than car batteries, and when they are plugged into the battery first, the capacitors in the charger fill up so quickly that they cause a spark. It therefore would be a better option to plug an electric bicycle charger into the main circuit first before plugging it into the battery. You usually can hear the spark when the wrong connection is made first, so it's best to connect whatever end first that doesn't cause a sparking noise. It also might say in the manual which end to connect first.

Avoid taking the charger with you on your electric bicycle because it may get damaged. If your charger is being bashed around inside a shoulder bag, components may come loose inside the case. It's best to have two chargers, one at the destination and one at home.

### 5.3.2 Hub Motor Maintenance Checklist

Very few things are needed to maintain a hub motor because they are very reliable.

## Tire Pressure

Check the tire pressure regularly. Take action if it is dropping, before you get caught out by a flat.

## Spokes

Check the spokes regularly; look out for snapped spokes or loose spoke nipples. Tighten and replace the spokes as needed. Spokes usually snap at the hub. It is sometimes possible to change a broken spoke without having to take off the wheel. Just unscrew the old one, leaving the nipple held in place by the inflated inner tube. Then thread the new spoke through the hub, and turn the nipple around and up onto the spoke. The spoke may need to be bent slightly to get it into position.

## Bearings

If you feel the wheel bearings break, then stop riding immediately; otherwise, you might cause damage to the Hall sensors or magnets. Change the bearings when they wear out (every 1 to 2 years) or whenever the hub motor is open for other maintenance. See Section 5.4.4 for how to open up a hub motor and Section 5.4.9 for how to change the bearings on a hub motor.

## Rims

Check the rims for damage, buckling, and weld damage or corrosion. If you are using rim brakes, then clean the rubber residue off the rims every few weeks.

## Connectors

Check for contact corrosion caused by sparking and melting caused by high current flow.

### 5.3.3 Battery Maintenance Checklist

Batteries are a consumable item and will yield a service life based on how well they are treated.

- *Always keep the battery topped up.*
  It is generally good for batteries to be fully charged, especially lead-acid batteries. Topping up lithium batteries also enables the BMS to balance the cells.
- *Check that the charger is putting out the correct voltage.*
  Sometimes charger voltage can be too low, and this won't activate the BMS balancing function at the end of battery charging. For example, a 48-V LiFePO$_4$ battery charger should put out 61.5 V.

- *Don't leave the battery unused for more than a month.*
  Lithium cells don't have a very fast self-discharge, usually around 1 percent a year. However, the BMS will drain the battery at a faster rate than this. A number of people have left their batteries over winter while not using their bicycles, only to come back to a nonfunctional pack in spring because the BMS drained the pack down to 0 V. To prevent this, it's best to either unplug the BMS or put the pack on its charger using a 7-day timer plug to keep it topped up for a short period once every week.
- *Don't overdischarge, short circuit, overheat, or drop your battery.*
- *Don't series-connect with different capacity or state-of-charge batteries.*
- *Don't parallel-connect with different voltage packs.*
- *Don't put things such as bicycle locks in the same container as the battery.*

### Thermal Management

The batteries shouldn't heat up much during operation; otherwise, there is something wrong. Sometimes it can be difficult to monitor temperature by touch if the battery is in a strong box. Monitor battery temperature on a new battery or new controller or when any component changes on the electric bicycle. Don't exceed 40°C in everyday use.

### Current Draw

Make sure that you do not exceed the peak or continuous current rating of the batteries. Monitor this at least on a new build or when any component changes on the bicycle, if not all the time.

### Voltage Drop ($V_{min}$)

It's easier to measure battery voltage than current. The voltage should not drop far below the nominal working voltage of the battery. Monitor this at least on a new build or when anything changes on the bicycle, if not all the time.

### Capacity

To test the battery, you might want to do a long ride to drain the battery and measure capacity and range of the bicycle. It's not good to 100 percent discharge the battery on a regular basis. It's good to do this when you first get the battery and then once every year to check that you still have the full capacity.

### Insulation and Padding

Check that the battery insulation and padding is not damaged. It can become damaged by attrition if the battery bounces around a lot. Repair the damage, and use a more secure battery mounting to prevent it from happening again.

**Corrosion on Cells**

Check for corrosion. Corrosion or a deposit around the cells usually means that either the cells are leaking electrolyte, have vented electrolyte at some point, or are being exposed to water.

### 5.3.4  Controller Maintenance Checklist

- Don't use lots of power when moving slowly or when stationary; this will blow the capacitors.
- Don't ride in sand or thick mud for long periods; this will overheat the controller.
- Make sure that the controller is not getting hot.
- Make sure that the throttle cable is protected from rain. Don't allow insulation tape to collect water around the throttle cables. Usually they are splash-proof to some extent.
- Make sure that none of the wires gets worn or trapped within the bicycle frame, suspension, or battery mounts.

### 5.3.5  Bicycle Maintenance

- Check brake pads once every 3 months for disc brakes and every month for rim brakes.
- Check brake adjustment once every 3 months for disc brakes and every 2 weeks for rim brakes.
- Check battery mounts for signs of bending or movement under the weight of the batteries once every 3 months.
- Check headset bearings for signs of excessive play.
- Check lights every month for damage or movement.

## 5.4  Repairing Electric Bicycle Components

### 5.4.1  How to Repair a Controller with Broken FETs

Broken FETs result in a fault that usually occurs in a controller operating near its voltage limit. This problem can be fixed, the replacement parts are not expensive, and it saves you the cost of buying a new controller. You don't always need to replace all the FETs, only the ones that are shorted out. The controller usually has groups of FETs ganged in parallel to increase the current-carrying capability (Figures 5.17 and 5.18). This means that the FETs short circuit in groups. Usually, a controller will have multiples of three FETs for the phases. Each phase usually will have two FETs for switching positive

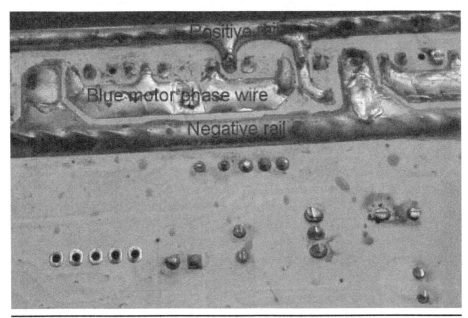

**FIGURE 5.17**    Underneath the circuit board, where the legs of the FETs come through. The big masses of solder are the main power rails and the motor phase rail.

**FIGURE 5.18**    The FETs. There are four FETs for each phase, two parallel to switch positive and two parallel to switch negative.

and two for switching negative. A controller will either contain 6, 12, or 18 FETs depending on how powerful it is. If only one short circuit is detected between phase wires, then you can open the controller and trace the broken group of FETs causing the short circuit by continuity tracing the wires back from the phase wire to the group of FETs.

There are lots of different FETs available with different current and voltage specifications. Therefore, when you buy the replacement, you may want to go for an upgrade at the same time because this will allow your controller to deliver more current and be more powerful. So you may need to decide that you just want to replace the broken FETs or upgrade all the FETs. The best-quality FETs are IR4110's. These can deliver 100 A and work at 100 V. Table 5.3 shows the other choice of FETs available and their current-carrying capability. When your replacement FETs arrive, it's a simple matter of desoldering the old FETs and soldering in the new ones. However, it is a complex soldering operation and requires a little skill and coordination with a

**TABLE 5.3**  FET Selection Guide

| Part | Max voltage | Resistance (mOhms) | Max amps |
|---|---|---|---|
| IRFB3077PBF | 75 | 3.3 | 210 |
| IRFB3207ZPBF | 75 | 4.1 | 170 |
| IRFB3207 | 75 | 4.5 | 180 |
| IRF2907Z | 75 | 4.5 | 170 |
| IRFB3307ZPBF | 75 | 5.8 | 120 |
| IRFB3307 | 75 | 6.3 | 130 |
| IRF3808 | 75 | 7 | 140 |
| IRF1607 | 75 | 7.5 | 142 |
| IRF1407 | 75 | 7.8 | 130 |
| IRFB3507 | 75 | 8.8 | 97 |
| IRFB4110PBF | 100 | 4.5 | 180 |
| IRFB4310 | 100 | 7 | 140 |
| BUK7510-100B | 100 | 8.6 | 110 |
| IRFB4410 | 100 | 10 | 96 |
| IRFB4710 | 100 | 14 | 75 |
| IRFB4610 | 100 | 14 | 73 |
| IRF8010 | 100 | 15 | 80 |
| IRF3710Z | 100 | 18 | 59 |
| IRF3710 | 100 | 23 | 57 |
| IRFB59N10D | 100 | 25 | 59 |
| IRF540Z | 100 | 26.5 | 36 |

solder sucker and a powerful >40-W soldering iron. Replacing FETs on a controller will be the same as replacing FETs on a BMS. Similar instructions will apply. Thanks to Richard Fechter in San Francisco for working out this repair procedure.

1. Open the controller.
2. Continuity trace the shorted phase back to the circuit board, and track down which FETs have short circuited. It may be all of them or just one phase group.
3. Decide to upgrade all the FETs or just replace the broken ones.
4. Order the FETs you need.
5. Unscrew the FETs from the heat sink.
6. Remove the plastic insulating sleeves and washers that insulate the FETs from the screws. Keep the screws and insulators safe.
7. Snip off the dead FETs at the legs, leaving behind as much leg as possible so that you can remove them later with pliers.
8. Now you need to remove the leg left behind and clear the holes of solder to insert the new FETs.
   • Debulk the mass of solder to make it easier to remove the FET legs (Figure 5.19). Using the soldering iron, heat up the mass of solder around the stumps left from the legs of the dead FETs. Don't leave

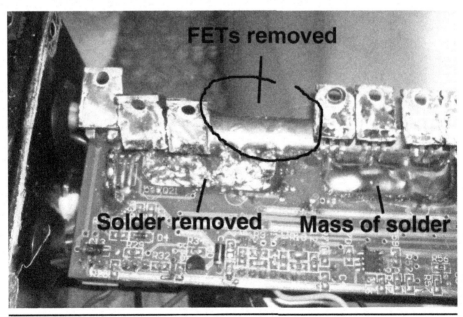

**FIGURE 5.19**  This shows the gap where the dead FETs have been removed. Debulk the mass of solder so that you can remove the dead FET legs.

the iron on the board for more than 8 seconds or you might lift a track from the surface of the board, which is difficult to replace. Suck up the solder using the solder sucker and iron.

- Remove the legs left from the dead FETs using tweezers and the soldering iron or the solder sucker. Gently pull on the stump while melting the solder around it. Don't use force, or it will lift a track on the board. It should come out easily when the solder is melted all the way through. Sometimes you can poke the leg out; sometimes you can solder suck it out. Use whatever it takes to remove the leg stumps, but don't lift a track. Be especially careful on the third leg, the one that is not attached to a great lump of solder. The traces here are easy to pull out. Use a lower temperature setting if you have a temperature-controlled soldering iron.
- Clear the holes left by the dead FETs using the solder sucker. Suck from one side of the board while heating from the other. Poking through with a wire while heating sometimes helps, too.

9. Bend the legs on the new FETs over in the same way as the other FETs.
10. Position the new FETs in the holes in the circuit board.
11. Screw all the FETs back onto the heat sink. It's best to screw the FETs into the heat sink first, as a guide for soldering, because if you do it the other way around, they might not line up with the heat sink properly.
12. Solder the new FETs into position. If you find it difficult soldering with the heat sink in the way, then just do one leg on each new FET to hold it in place, and then remove the heat sink and solder all the other legs properly. Be careful not to accidentally connect neighboring traces or legs with solder.
13. Add loads of extra solder back to the main busbars to beef up the current-carrying capability.
14. Add extra thermal paste to the FETs, and then fix them onto the heat sink for the last time. Remember to use the insulating sleeves and washers on the screws.
15. If you removed a trace during this process, follow where it went, and solder a jumper lead to repair it. If it was on the third leg, then it will lead to the gate resistor. You can solder a jumper lead from the jumper resistor.
16. Measure the resistance again between all the phases and each power wire. Make sure that there is no short circuit.
17. Check that the gate resistors are working properly (Figure 5.20). When a FET blows, it sometimes takes out its gate resistor. Use a multimeter to resistance test between the third pins on each set of FETs. This should read 20 $\Omega$. Each gate resistor is 10 $\Omega$. The resistors blow open circuit, so if there is a problem, the resistance will be much higher than 10 $\Omega$. The gate resistors are shown in Figure 5.20 and are usually located close to

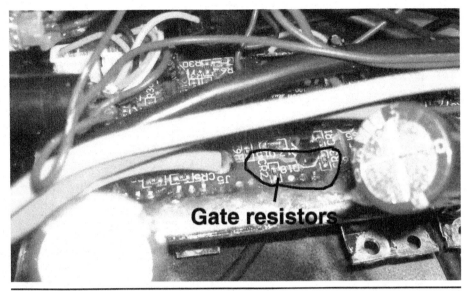

**FIGURE 5.20** Here, the gate resistors are visible. These may need to be replaced as well. The gate resistors are connected to the third leg of the FETs.

the FETs that blew. You can continuity check to make sure because it will be connected to the third leg of the FET. If the gate resistor blew, then chances are that the gate drive transistor also blew, and these are difficult to test.

18. Replace any blown resistors with new ones.
19. Check the board for stray pieces of solder that could cause short circuits.
20. If there are big deposits of solder flux around the traces, then carefully scratch them away with a small screwdriver. Solder flux can be slightly conductive and could cause problems.
21. When finished (Figure 5.21), put the controller back into its 11 enclosure, and test it on the bicycle. For testing, it is best to use a current-limited supply. You can do this by putting a light bulb in series with the battery (e.g., slot an MR16 10-W bulb into the fuse holder). If it still doesn't work and you are sure that you did everything right, then probably the gate driver chip has blown too.
22. If the test is successful, then put the controller back on the bicycle and clean up. If the test was not successful, then replace the gate driver transistor as in the next section.

### 5.4.2  Repairing a Controller with Broken Gate Drive Transistor

When you replace the FETs on a controller you sometimes have to replace the gate driver transistor as well. If a gate resistor blew, then chances are

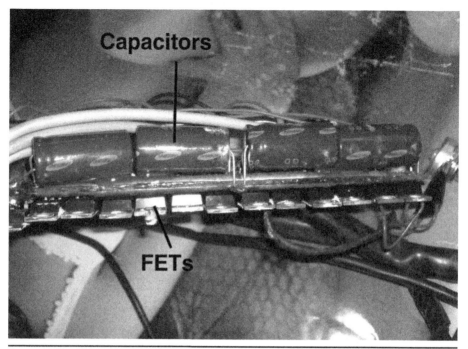

**FIGURE 5.21**    This side view shows the capacitors and all the FETs.

that this is the case. Replacing the gate drive transistor is a very difficult soldering job because it is surface-mount technology. This repair will require two soldering irons, a fine-tipped iron and a temperature-controlled iron.

1.  Locate the gate driver chip. The boards are usually split into thirds for each of the three phases, so it will be near the FETs that blew. The most common gate driver chip is the IR2101. This is what is used on the Crystalyte V1 and V2 controllers.
2.  Buy a replacement IR2101. They come in both surface-mount and chip forms. The surface-mount form is easiest to use.
3.  Remove the dead chip. Use two soldering irons at medium heat, not tinned, and position them on the legs on either side of the surface-mount chip. Wait until the solder melts, holding it to the board, and then move it away from the solder pads with the irons. Move it to an uncluttered area of the board, and then retract the soldering irons. Hopefully, the chip won't stick to something else on the board as the solder sets. Remove it when it is free.
4.  Make sure that none of the solder pads are shorted. Clean up the solder pads by gently melting the solder and sweeping it with the iron.
5.  Solder the new chip in place.

- If you have a surface-mount chip, then position it on the solder pads, and heat the legs individually with the iron. A piece of wire wrapped round the soldering iron called a *stinger* can help with soldering the little legs of the surface-mount chip. Be careful not to short circuit the legs; this is delicate work. Use a magnifying glass if needed.
- If you have a through-hole chip, then you will have to fashion something like the spider shown in Figure 5.22. Solder wires to the solder pads first and then to the chip legs.

6. Continuity check between neighboring legs to check for short circuits.
7. Put everything back inside the case. If you did the spider technique as shown in Figure 5.22, then you must insulate the chip with electrical tape and secure it to the other wires inside the case to stop it from coming loose.

**FIGURE 5.22** Replacement IR2101 gate driver chip. Also visible is the variable-current mod.

### 5.4.3 How to Repair a Controller with Blown Capacitors

Replacing blown capacitors is easier than replacing FETs. The hardest part is fitting the new capacitors inside the case because they can be very large. If you blow the capacitors more than once, then you may be working too close to their voltage limit, and it may be worth upgrading them to a higher voltage. Blown capacitors usually are caused by very low-speed, high-power-demand conditions, so you might consider changing the gearing of the motor if it happens again. Blowing up also can occur because they were low-quality or aging capacitors. As capacitors age, their equivalent series resistance (ESR) increases, they heat up, and then they explode. It's best to replace them with bigger capacitors with lower ESR or more capacitors. See Figures 5.23 through 5.26.

**FIGURE 5.23**   Capacitors are just connected across the main negative and positive supply rails, as shown from underneath.

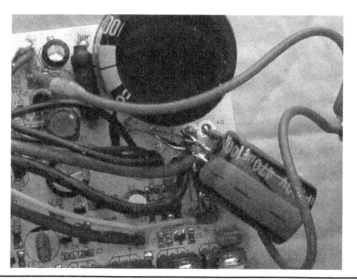

**FIGURE 5.24**   Replacement of one capacitor—same capacity, same or higher voltage.

**FIGURE 5.25**   Note the burned-out trace repaired by careful placement of a blob of solder to join the broken ends.

**FIGURE 5.26**   Replacement of the other capacitor. Watch out, they can still fit back in the case.

1. Snip off the dead capacitor, and pick off any debris from the circuit board.
2. Look to see if there are any melted traces on the board that need repairing (see Figure 5.25).

3. If there are melted traces, then you will need to solder in a makeshift jumper wire to replace the broken trace. The jumper(s) wire should connect the places that are no longer connected by the broken trace. This may require some skillful soldering, but if much of the board is destroyed, the damage may be irreparable.
4. Next, snip off the dead capacitors, leaving as much leg behind as possible.
5. Using the soldering iron, heat up the mass of solder around the stumps left from the legs of the dead capacitors. Don't leave the iron on the board for more than 8 seconds, or you might lift a track from the surface of the board, which is difficult to replace.
6. Suck up the solder using the solder sucker and iron.
7. Remove the legs left from the dead capacitors using tweezers and the soldering iron. Gently pull on the stump while melting the solder around it. Don't use force, or it will lift a track. It should come out easily when the solder is melted all the way through.
8. Clear the holes left by the dead capacitors using the solder sucker. Suck from one side of the board while heating from the other.
9. Find the best location for the capacitors in the board. Make sure that the capacitors don't prevent the case from sliding back on to the board (see Figure 5.26). Make sure that no traces touch the case. The capacitors can go anywhere in the case where there is space to fit them, but it's best to keep the wires short.
10. Put the capacitor legs through the holes, and solder them in position
11. Add loads of extra solder back to the main busbars to beef up the current-carrying capability.
12. Glue down the capacitors to stop them from vibrating.
13. Close the case and finish.

## 5.4.4 Opening Up a Brushless, Nongeared Hub Motor

The brushless gearless motor is the simplest and easiest to disassemble. It has only four main parts. The stator is the big wire-wound centerpiece in Figure 5.29. This is fixed to the axle and is stationary. The wheel housing hub with the magnets is around the outside of this part, and the cover plates all rotate. See Figures 5.27 through 5.29.

1. Remove both side covers from the motor. It may be slightly easier to remove the gear cassette from the right side cover before removing the cover. You need a gear puller to remove the gear cassette.
   - Fit the gear puller, and use a wrench on it to unscrew the gear cassette.
   - Gently unscrew all the bolts holding the side covers in place. Be careful not to shear off the bolts.

**FIGURE 5.27**    Inside, the nongeared brushless hub motor is very simple. Here is where the wires enter the hub. The inside (stationary part) is the stator; the outside (rotating) part is the magnet ring.

**FIGURE 5.28**    The other side looks the same, but you can't see the wires entering the hub.

**FIGURE 5.29**   Here is the stator removed from the magnet ring.

- Carefully pry open the side covers evenly all along the seam with a knife or screwdriver.
- The covers should pop out of the wheel housing as the magnet ring comes off the ring in the side cover.
- Then there should be enough space to get your fingers behind the cover and pull it off. Pull the cover and bearing off the axle until the side cover is free. Do this to both side covers.

2. Next, get the stator out of the hub magnets. The stator is the middle part attached to the axle. The magnets are very strong, so you need a lot of force to achieve this. Be careful not to trap your fingers because the magnets can pull on the stator very quickly. Also be careful not to fall over while you do this.

- Put the coverless hub motor on hard ground with the axle and rim touching. If you remove only one side cover, then the open side of the motor must face up.
- Step on one side of the wheel rim while the axle contacts the ground.
- The opposite edge of the wheel will be sticking in the air. To get the stator out, step on the other side of the wheel and use your weight to push down on both sides of the wheel while the ground holds the stator. It should pop out as soon as you get away from the action of the magnets.
- Take the stator away from the magnets.

### 5.4.5  Reassembly of the Brushless, Nongeared Hub Motor

1. Put the stator back into the hub. Be careful that you don't get your fingers damaged while you do this because the magnets are extremely strong. Putting the stator back in is basically the reverse of step 2 above, taking it out.
   - Put the cover-less hub wheel on the ground, and stand on opposite sides of the rim.
   - Hold the stator in the middle with the axle touching the ground.
   - Slowly lift the rim into the stator while using your weight to prevent violent movement by the magnets that may cause damage. The stator will pop back into position between the magnets.
2. Put the side covers back on.
   - Slide the side cover bearing back onto the axle.
   - The stator will be off center, attracted by the magnets. Center the axle by slotting the magnet ring over the ring in the side covers.
3. Put the bolts back into the side covers. It may be difficult to tighten the bolts because tension in the spokes can stretch the magnet ring out of alignment with the cover plate bolt holes. You may have to loosen all the spokes in order to match up the holes in the cover plate with the holes in the magnet ring. You need a spoke wrench to loosen the spokes.
4. Fasten the bolts. Be careful not to shear off the bolts when tightening them.
5. Then retension the spokes.

### 5.4.6  Opening Up a Brushed Hub Motor

The brushed hub motor is a little more complicated to disassemble because the brushes are spring-loaded. It consists of four main parts: the hub-mounted stator that rotates, the side covers that rotate, and the magnet ring mounted on the axle with the brushes, which is stationary. In this respect, it is opposite to the brushless motor in that the positions of the stator and magnets are different.

Remove both side covers from the motor. It may be slightly easier to remove the gear cassette from the right side cover before removing the cover. You need a gear puller to remove the gear cassette.

1. Fit the gear puller, and use a wrench on it to unscrew the gear cassette.
2. Gently unscrew all the bolts holding the side covers in place. Be careful not to shear off the bolts.
3. Carefully pry open the side covers (Figure 5.30) evenly all along the seam with a knife or screwdriver. Be careful not to slip and damage the windings. Sometimes the windings are protected by a plastic shield for this reason.

4.  The covers should pop out of the wheel housing as the magnet ring comes off the ring in the side cover.

5.  Then there should be enough space to get your fingers behind the cover and pull it off. Pull the cover and bearing off the axle until the side cover is free. Do this to both side covers. Sometimes they are stiff and require extra force or leverage. When the covers are off, it should look something like Figures 5.31 through 5.33. At this point, you can access the bearings.

**FIGURE 5.30**   Lever the side covers off. Be very careful not to slip and damage the windings.

1.  On the brushed motor, it is not advisable to remove the stator from the magnets because the brushes are spring-loaded and are very tricky to get back in during reassembly. It is also difficult to find replacement brushes. On the brushed motor, the magnets are in the center, as shown in Figure 5.32. The magnet ring can only be taken out of the hub on one side.

2.  To remove the magnet ring, use the "step on" technique described earlier. Make sure that you have the yellow brush-holding plate side in Figure 5.31 facing the ground and the magnet side facing up. Step on one side of the wheel rim with the axle touching the ground. Then step on the other side to push the inner magnet ring and axle assembly out of the hub.

3.  Reassembly is difficult and may require you to tape the brushes into the holders while you put the magnet ring back into the stator.

**FIGURE 5.31**   This side shows the windings and the rear of the contractors. The bearing is visible in the center of the cover plate. Holes have been drilled in this plate for heat dissipation.

**FIGURE 5.32**   This side shows the magnets attached to the axle and the connections to the brushes. The bearing is visible in the cover plate.

**FIGURE 5.33**   This view shows the inner workings. One of the two brushes is visible. The brush skims over a disk of flat copper contacts on the other side of the yellow board. This is what does the commutation in the brushed motor.

4. Use the "step off" method very carefully to ensure that the magnet ring doesn't hit the contactor plate with force because this might cause damage.
5. Take the tape off the brushes using tweezers from the other side.
6. Put the cover plates back on.
7. Screw the bolts back into the cover plates to secure them.

### 5.4.7 Opening Up a Brushless, Geared Hub Motor

The geared brushless motor has only one side cover, and the motor windings are hidden inside an inner motor compartment. This motor has many gears and moving parts and is more complicated than the nongeared types. Geared motors are not supposed to be completely disassembled.

1. Unscrew the hub bolts.
2. Take off the side cover (Figure 5.34).
3. From the other side, hit the axle with something to push the motor unit out of the hub (Figure 5.35).
4. Now, both bearings are accessible (Figure 5.36).

**FIGURE 5.34**    Undo the bolts and lever the side cover off.

**FIGURE 5.35**    Hit the axle from the other side with something, and the inner motor will pop out of the hub.

5. To access the windings and Hall sensors, open the back of the motor unit. Undo the bolts, and it should pop open (Figure 5.37).

6. Reassembly is just the reverse of this disassembly.

**FIGURE 5.36**    Here you can see the hub bearing and the planetary gears on the motor; the other bearing is in the side cover.

**FIGURE 5.37**    When you open the other end of the inner motor, you access the Hall sensors, windings, and magnets.

### 5.4.8  Replacing Hall Sensors in a Brushless, Nongeared Hub Motor

The Hall sensors usually need replacing if you spin the motor without securing it properly and wrap the wires around the axle. Sometimes corrosion can cause broken Halls, but usually they are coated in silicone, so this is rare. Usually it is the wire that breaks, and you should eliminate this possibility first before you open up the hub motor. The Hall sensors you need are the bipolar latching type with an open collector output, something such as SS40A, SS41, or SS466A.

1. Open up the motor as described in the preceding section.
2. Examine the Hall sensors for damage. Look for broken wires or corrosion. Find the sensor(s) that need replacing.
3. The Hall switching test in Section 5.2.3 can be used to decide which Hall sensors are broken. You don't have to replace all the Hall sensors if only one is broken. Follow the wiring color of the broken, nonswitching sense wires back to the motor to find the broken Hall sensors.
4. Remove the broken Hall sensors any way that is possible (Figure 5.38). It is sometimes possible to melt the hot glue holding a sensor in position and pick it out. Other times you may have to destroy it to get it out. Clean any debris from the notch in the stator.

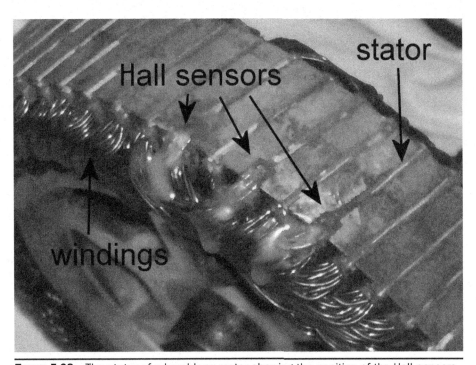

**FIGURE 5.38**  The stator of a brushless motor showing the position of the Hall sensors.

5.  Connect the new Hall sensor. This is a delicate job, and it's easier to solder wires onto the sensor first and then solder those wires onto the motor leads. Each Hall sensor should have a positive supply (red), negative supply (black), and a sense wire.
6.  Now glue the Hall sensors into the notches where the old sensors were. Don't get the positions of the sensors and their colors mixed up.
7.  Make sure that all the wires are shielded from the inside walls of the rotating hub. Hot glue them in place, if necessary.
8.  Reassemble the hub as described in the preceding section.

### 5.4.9  Replacing the Bearings in a Hub Motor

The bearings wear out in about a year of use on some of the big hub motors. It is important to know how to change them because your local bicycle shop won't know. The replacement bearings may be of different sizes for each side of the hub. On the X5 motor, the bearings are 6005 on the left and 6004 on the right. If you don't know what size bearings to use, there are three measurements you have to take before you can order a replacements: the inside diameter, the outside diameter, and the thickness. Make these measurements with a vernier caliper on the shaft, cover plate hole diameter, and thickness. The bearings usually have the part number on them, but they are sometimes destroyed too badly to read.

1.  Remove the gear cassette from the hub. You need a gear puller to remove the gear cassette. Fit the gear puller, and use a wrench on it to unscrew the gear cassette.
2.  Remove both side covers from the motor.
    * Unscrew all the bolts holding the side covers in place. Be careful not to shear off the bolts.
    * Carefully pry open the side covers evenly all along the seam with a knife or screwdriver.
    * The covers should pop out of the wheel housing as the magnets pull the axle off center.
    * Then there should be enough space to get your fingers behind the cover and pull it off. Pull the cover and bearing off the axle until the side cover is free. Do this to both side covers.
3.  The best way to get the bearings out is to press them out. If you have access to a hand press, use it. If you don't, then it is possible to tap the bearings out. If the bearing is badly damaged or disintegrated, then you may need to lever and dig it out piece by piece.
4.  Hand press method:
    * Position the side plate on the hand press with the bearing side facing down.

- Put something flat and disk-shaped (such as a coin) inside the axle hole resting on the rim of the bearing.
- Press the bearing out using the hand press.

5. Tapping method:
- Position the side plate on some blocks with the bearing side facing down.
- Put something flat and disk-shaped inside the axle hole resting on the rim of the bearing.
- Use a pipe like a chisel to tap on the disk to push the bearing out.
- Tap gently and evenly around the disk, and keep it parallel with the side cover.

6. To fit the new bearing, just reverse the procedure, tapping or pressing from the other side. Use grease to help ease the bearing into position.

### 5.4.10 Repairing a Battery with Dead Cells

#### Nickel Battery Without a BMS

Nickel batteries tend not to have very sophisticated BMSs. They don't need them for normal operation. This allows a simple maintenance trick. If a single cell in an NiMh battery breaks open circuit, then there is a very easy solution. The cell can be short circuited, and the battery will go on working. The charger will see a reduced voltage, but typically NiMh chargers work based on constant current and a temperature cutoff limit that will be unaffected. Make sure that the cell is dead before you short circuit it; otherwise, there could be smoke. If it shows 0 V at the terminals and there is no conductivity through it, then it is broken open circuit, and it is okay to fit a bypass wire. The bypass wire should be thick enough to carry main power current.

#### Replacing a Dead Cell that Has a BMS or Constant-Voltage Charger

Usually, for one of many reasons, it is necessary to replace a dead cell. If the battery has a BMS, if the pack is to be paralleled with another pack, or if the battery charger is of the constant-voltage type, then it is necessary to replace a dead cell. To replace a cell in a battery, you have to locate the dead cell, remove it, and then fix in a replacement. The way you do this will depend on the configuration, how easy it is to get at the cell, and how the cells are joined. If the cell is very difficult to get at, then it may be best to solder a short across the dead cell and then fix a new cell into the end of the pack, where it is easier to do. This would require a rearrangement of the balance wires if the pack has a BMS (see Section 5.4.11). Replacing a dead cell in a pack may require complete disassembly and reassembly of the pack. If this is the case, then read Section 6.1.1. When working on batteries, avoid using metal tools whenever possible; work on a clean, empty, nonconductive surface; and don't wear jewelry. Remember there is no off switch on a battery you have to work on them live.

### Replacing a Dead Cell in a Pack

1. Disassemble the pack covering. Cut open the heat-shrink or duct-tape shell. Be careful not to stab or cut any of the cells; they may only be a few millimeters under the surface of the packaging.
2. Measure the cell voltages to locate the dead cell.
3. Cut the dead cell out at the contact tabs. Use side cutters to cut the metal tabs joining the dead cell to its neighbors.
4. Remove the cell. Break off any hot glue used to secure the cell, and push it out.
5. Position a new cell with contact tabs.
6. Solder the contact tabs to the correct cells (see "Soldering to Batteries" in Section 6.1.1).
7. Refit the cell insulation.
8. Re-cover the battery pack (heat shrink or duct tape).
8. Check that the pack still charges.

## 5.4.11 Fixing Broken Balance Wires in a Pack That Has a BMS

If a battery pack with a BMS is knocked around a lot or there was insufficient strain relief on the balance wires, then this can result in a broken balance wire. A broken balance wire will cause the BMS to trip and stop you dead in your tracks. To replace the balance wire, you have to open up the pack to find the broken wire and then fix it back in place. When working on batteries, avoid using metal tools whenever possible; work on a clean, empty, nonconductive surface; and don't wear jewelry.

1. Disassemble the pack covering. Cut open the heat-shrink or duct-tape shell. If the cells are pouch cells, be ultracareful not to stab or cut any of the soft cells; they may be a only few millimeters under the surface of the packaging.
2. Check all the cell voltages at the cell terminals.
3. Look for the broken BMS wire (Figure 5.39).
4. If several balance wires are disconnected, don't guess where to connect them. Find out where each wire was supposed to be connected by tracing the continuity of the wire back to the position on the BMS. The sequence of the balance wires in the BMS will tell you which wire needs to go to which cell terminal. They will be ordered from positive to negative.
5. Solder the wire back in place.

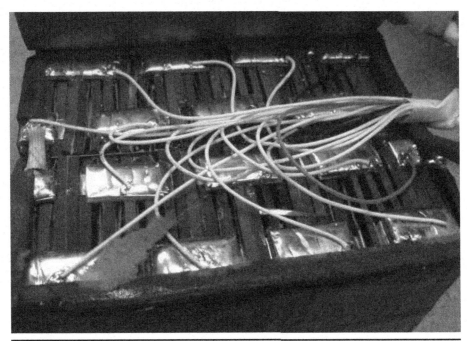

**FIGURE 5.39**   Balance wires connecting the BMS to cell terminals.

6. If any of the balance wires look too short or under tension, as in Figure 5.39, then extend them with extra wire of similar type.
7. While your pack is open, inspect the cell insulation and padding. Check that it is sufficient for the wear and tear that you are putting on it.
8. Close up the pack, and reapply the heat shrink or duct tape.

### 5.4.12  Repairing a Battery Charger

The usual causes of battery charger failure are overheating, voltage spikes, and rough handling by the user. It is rarely possible to repair a broken bicycle charger. If there is an obvious blown component to swap that you can identify and replace, then replacing this component might fix the charger. Capacitors are one of the most likely components to blow through voltage spikes. I've had several bicycle chargers blow and not been able to fix them (Figure 5.40) despite changing capacitors, resistors, and transistors. Prevention is better than a cure, so follow the mods in the maintenance section if you think the charger runs hot, etc. If your charger does fail, the best thing to do is to buy a different brand of the same voltage. Make sure that you follow the wiring guide for a new battery pack or charger in Section 5.4.13 so that you get the connections correct. They might not be the same as the old charger. Make sure that the charger is disconnected from the main circuit before you work on it.

**FIGURE 5.40**   A cheap NiMh battery charger (prone to breaking after 6 months).

### 5.4.13  Wiring Up a New Battery Pack or Charger

When you buy an electric bicycle package, everything is set up for you, but when you buy a new battery or charger to work with an existing kit, then there may be some minor but important wiring jobs for you to do. It is important not to rush these and to know what you are doing because it is possible to wreck these lovely new toys with a single ill-thought connection.

#### NiMh Pack and Charger Connections

Most NiMh packs will come with a thermistor already attached. Some battery chargers are different, though, and use different value thermistors, so it is important to check that the new pack will work with the charger you have. Likewise, it is important to check that a new charger operates using the same thermistor resistance as your old one. If you don't check the thermistor value, then the pack might not charge, or worse, it might not recognize when the charge has finished and overcharge your new battery. The thermistor is usually connected in the battery across the negative and third terminal of the plug, as shown in the circuit diagram in Figure 5.41. Using the multimeter on voltage setting, check all the pins, and find which are positive and negative. The thermistor is connected to negative, so it also will have a voltage.

It will appear as if there are two negative pins. The one with the lowest voltage with respect to positive will be the true negative; the other one is connected to the thermistor. Using resistance measurement, check the resistance between the negative and sensor pins. It should be between 1 and 20 kΩ. Now do the same checks and measurements on the battery pack that

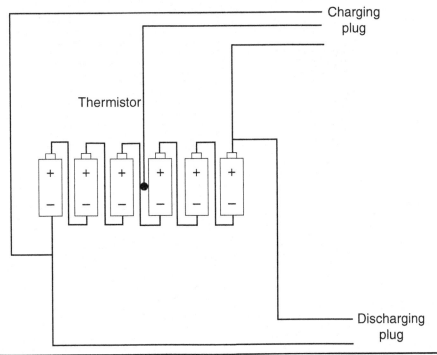

**FIGURE 5.41**  How to connect an NiMh pack showing the thermistor and the charger plug.

came with the charger. Make sure that both packs are the same temperature, or the readings will be different. You want to make the resistor values the same. If they are not, then you will have to open up and change the thermistor in the new pack to match the value the charger expects. Thermistors are rated by their highest resistance, so when you are buying a replacement thermistor based on the resistance reading, round up the value from when the pack was coldest, and buy a thermistor close to that resistance, or check the resistance with a meter before you buy.

**Lithium Pack with BMS Connections**

Lithium packs usually need to have a BMS to balance them and ensure that no cell drops below a minimum voltage limit. The BMS is connected to the main negative on the pack and then offers to the user separate controlled negative connections for charging and discharging (Figure 5.42). The BMS can electrically switch off the connection to these negatives if it detects that something is wrong with a cell. The BMS also has a connection to the battery positive, from which it draws operating power, and connection taps between each cell. Usually there is no direct connection to battery negative for the user, but it can be useful to have one for fault finding or emergency pur-

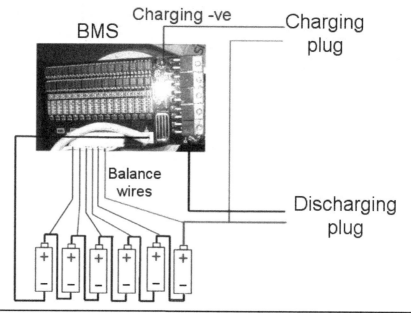

**Figure 5.42** How to connect a lithium pack with a BMS including charger port and balance wires.

poses. The main BMS connections should be clearly marked on the circuit board by + and – symbols. The BMS wires should be clearly marked with "discharging" and "charging" labels. The negative connections also should be indicated on the board if the BMS comes without negative wires attached. If you buy a large battery and need to separate it into smaller sections to fit on your electric bicycle, then you have to be mindful of BMS connections. Some packs have series and parallel cells, so it's possible to separate them in two ways. You should separate the pack the easiest way possible. If you separate a series string of cells, then the connecting main current wire between the two halves needs to be very thick. This is because any voltage drop in this wire will be assumed by the BMS to be a result of a bad cell. Any Anderson connectors or fuses here could cause BMS cutoffs. Always remember to reconnect the balance wires. If you separate the battery into two parallel halves, then you can use Figure 6.1 as a template.

### Connecting the Battery Plug and Charger

Sometimes when you buy a new battery, it will come with the charger plug removed, for you to connect yourself. This is done so that you can panel-mount the plug on your battery box and thread the wires through. Sometimes the wires will be clearly marked so that they can be connected easily. If they are not marked, then use your multimeter on voltage setting to check the polarity of the wires and compare that with the polarity of the charger when plugged in. The battery should be connected to the charger positive to positive and negative to negative. Polarity detection is explained in the tips in Section 5.1.1. If the BMS has separate negative wires for discharge and charge, then the smaller-thickness blue wire is usually the charge wire. The charge plug will have the negative BMS charge wire and a connection to positive. The discharge power wires that connect to the controller will be the thick positive wire and the thick black negative discharge wire.

# CHAPTER 6

# Projects for Electric Bicycles

## 6.1 Performance Modifications

### 6.1.1 How to Build a Battery Pack

It is very useful to know how to build a battery pack from spare cells for several reasons: It allows you to tailor-make packs to perfectly fit your electric bicycle project. You can use any cells you want, cells that may be free or cheap for whatever reason. Figure 6.1 shows a LiFePO$_4$ battery pack I made from used A123 cells designed to add a boost to my main battery pack and fit in a small empty space in my battery box. At the time of this writing, A123 cells are not available for the general public to buy. The cells I used were taken from power tool packs that developed faults. These are the good cells left over and thus were cheap. Both packs are based on LiFePO$_4$ chemistry, so I could use the same battery management system (BMS) to protect both packs. The same pack-building method can be used for any cylindrical cells, though.

Be very careful when working on batteries. The batteries on electric bicycles contain a lot of energy and can be very dangerous in untrained hands. The risk is highest when you are working on an open battery. When working on batteries, avoid using metal tools whenever possible; work on a clean, empty, nonconductive surface, and don't wear jewelry. Read the safety section (Section 1.3) and follow all the guidance. As the author of this book, I cannot be held responsible for any accident, injury, or damage to property caused by your work following instructions provided in this book.

The process of pack building is similar for all cell types, but there are some major differences that will be explained here. There are several methods of stacking cells to form a pack. The method you choose will depend on the type of cells you are using and the size of the pack you wish to create.

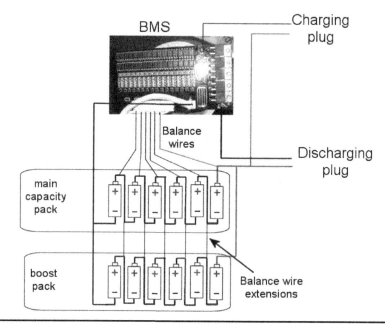

FIGURE 6.1    Adding cells to a lithium battery with a BMS so that the BMS controls both packs. This can be used to add high-rate cells to a high-capacity pack as long as they are the same chemistry and voltage.

When using cylindrical cells, there are two stacking methods that can be used for pack building: square stacking or triangular stacking (Figure 6.2). The figure shows a square stack of cells, where the cells line up in both directions and are not staggered between layers. This leaves lots of space between the cells, which is bad for volumetric energy density but easy to insulate. In a triangular stack configuration, the layers of cells are staggered like the bricks in a wall to fill the space between the cells more efficiently. For the type of pack in the figure, square stacking was chosen because the space in the battery box was very small, and the void space was important. In larger battery packs, it's important to stack the cells triangularly to achieve better volumetric cell density. There is only one way to stack flat-pouch cells. Because the cells are rectangular shaped, there is no wasted space between the cells when you square stack them. There is no benefit to staggering the layers of cells, so a triangular configuration is not used.

The next thing to decide is how to interconnect the cells. With cylindrical cells, you can either zigzag the cells and use straight battery tabs, or you can keep the cells lined up the same direction and leapfrog the connecting battery tabs (Figure 6.3). Usually, battery packs for electric bicycles use the zigzag configuration because it minimizes the length needed for the interconnecting battery tabs, which reduces the resistance and thus pack heating.

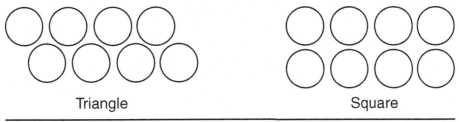

Triangle                                              Square

**FIGURE 6.2**    Triangular or square configuration of cells into a pack. The triangular configuration saves space in larger battery packs.

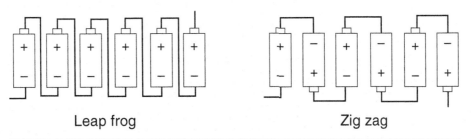

Leap frog                                              Zig zag

**FIGURE 6.3**    Leapfrog or zigzag connection of cell tabs in a pack. The zigzag arrangement
has shorter cell connections and therefore less pack heating.

The other thing that can differ in building a battery pack is whether you
join the cells together physically before joining them electrically or vice
versa. In the following example, the cells were joined electrically first,
through welding, and then they were joined physically by heat shrinking.
This allows the builder to add insulation between the cells as the pack is
constructed. Good pack insulation was a main priority because the insulation on the cells was poor. Usually NiMh battery packs are built the other
way around—by joining the cells first with hot glue and then welding them
together electrically afterward. This makes the pack look neater. It's best to
use circuit board or high-density card as insulation between the layers of
cells because such materials won't wear out.

Choose your battery tabs wisely for the current you are going to use.
Nickel, copper, and aluminum are good low-resistance battery tabs, but steel
is not. The resistance and voltage drop across the battery tabs can be calculated for your battery using the resistivity equation and Ohm's law (basic
law of resistance in conductors). First, look up the resistivity of the material
you are using for battery tabs on the Internet or in Table 6.1. Use Equation
6.1, the resistivity equation, to calculate the resistance of the tabs. Then use
Ohm's law (Equations 6.2 and 6.3) to calculate the voltage drop and power
consumed by the tabs at whatever current level you wish to operate the
pack. Aim to keep the wasted power below 10 W for a normal-sized battery
pack at the peak power you will use because more heat than this will not be

easy to dissipate and will increase the running temperature of the pack, which will shorten its life.

---

**EQUATION 6.1**

Resistivity × Total tab length / Tab cross-sectional area = Resistance

---

**TABLE 6.1**    Resistivity of Some Common Battery Materials

Choose the ones with the lowest resistivity.

| Material | Resistivity, $\Omega \bullet m$ ($\Omega \bullet in$) |
|---|---|
| Nickel | $6.99 \times 10^{-8}$ $(27.9 \times 10^{-6})$ |
| Copper | $1.72 \times 10^{-8}$ $(68.8 \times 10^{-8})$ |
| Stainless steel | $7.496 \times 10^{-7}$ $(2.99 \times 10^{-6})$ |
| Aluminium | $2.82 \times 10^{-8}$ $(1.12 \times 10^{-6})$ |

---

**EQUATION 6.2**    Ohm's law applied to current through battery tabs

Voltage drop across tabs = Current × Resistance

---

**EQUATION 6.3**    Energy dissipation in the battery tabs

$Current^2$ × Resistance = Power wasted heating battery tabs

Follow the instructions in Figures 6.4 through 6.21 to see how to build a battery pack. In Figure 6.5 notice how the tab is trimmed at the end to protect the cell insulation from abrasion and short circuit between positive and negative at the top sides of the can, where the positive and negative tabs are closest. Also notice the black insulating tape on the other side of the tab to protect from short circuits where the tab exits the cell and crosses the positive-negative gap. In a factory-made pack, this insulation would be heat-resistant. The most catastrophic failure occurs when there is heat buildup inside the battery that melts or burns the insulation and causes all the cells to short circuit.

To weld metal cylindrical cells together, you need to have a capacitance discharge welder, which is a specialist tool (Figure 6.6). The capacitance discharge (CD) welder works on the principle of a spot welder in that it doesn't heat up the whole cell, just the point that is to be welded. To weld with a CD welder, position the tab on the cell terminal. Then press both the elec-

**FIGURE 6.4**    Start off with two cells arranged end to end like this.

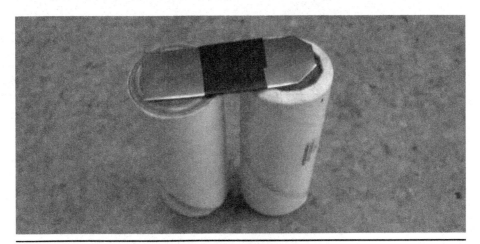

**FIGURE 6.5**    Cut, insulate, and position the first battery tab.

trodes down onto the tab, but not touching each other. Pressing on the cell in the place that you want to weld, you then depress the foot pedal, and the weld will be made (Figure 6.7). You will see some sparks, so wear goggles and protective clothing. You can vary the amount of contact force to change the intensity of the weld. A light touch of one electrode will ensure that all the power is delivered across the resistance caused by that contact and will create more sparks as a result. A hard-pressed electrode will produce a mild weld and is best for welding fragile things such as small balance wires. Try not to damage the cells, and avoid welding through insulation because the insulation will melt and cause heat and smoke. Welds should be done with welder tips always on the same battery terminal. Never make a weld across the two cell terminals, or the high current will destroy the cell. Now follow the directions in Figures 6.8 through 6.21 to complete your pack.

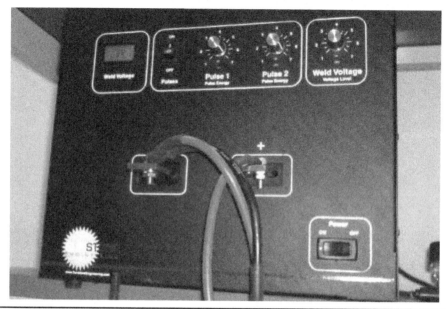

**FIGURE 6.6**   A capacitance discharge welder is the best way to join cells together.

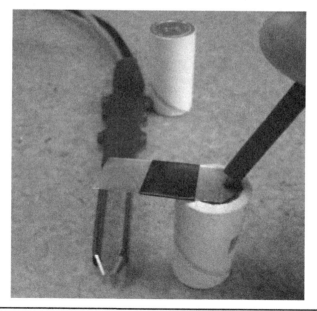

**FIGURE 6.7**   Position both electrodes on the tab, and depress the foot pedal. Never weld through the battery or it will explode! The weld electrodes always should be on the same battery terminal.

**FIGURE 6.8**    Do lots of welds to hold the tab on securely, but don't damage the cell with too much power.

**FIGURE 6.9**    The two cells are now securely joined both physically and electrically.

**FIGURE 6.10**    Weld the balance wire on too. This will require a lower setting and the right amount of tip pressure.

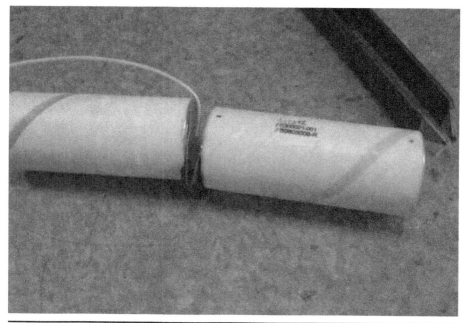

**FIGURE 6.11**    If the cell will be in the middle of the pack, then bend the tab and heat-shrink the cells to insulate them.

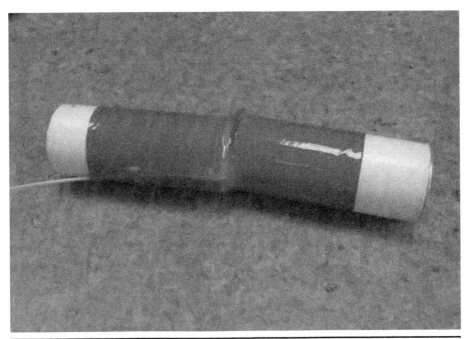

FIGURE 6.12   Good insulation is the most important thing in a battery pack. These two cells now become one with no exposed middle junction.

FIGURE 6.13   To make the ends of your battery pack, it is just like joining two cells. Remember to put a balance wire at every cell junction.

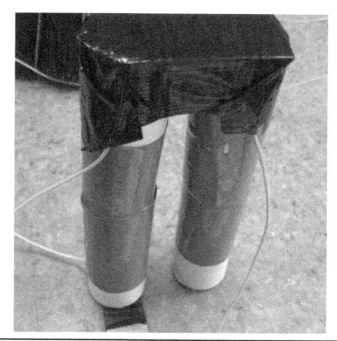

**FIGURE 6.14**    Insulate the join. This time duct tape and cardboard are used.

**FIGURE 6.15**    The battery pack is built up with lots of units in this way. Cardboard is used to separate the vertical columns of cells. Be careful of short circuits; insulate everything.

**FIGURE 6.16**    Here, a thin copper strip has been welded onto the terminal to make a good place on which to solder.

**FIGURE 6.17**    The almost-finished battery pack—16 cells in series, a 48-V pack.

Insulate the pack as you build. Use battery heat shrink to insulate the junctions of straight cell joined in the middle of the pack, and use duct tape and cardboard to insulate the cell junctions at the edges of the pack. Use cardboard or circuit board to insulate between cell layers. When the pack is completed, heat-shrink the whole pack.

**Figure 6.18**   With larger packs, keeping things straight can be a problem.

**FIGURE 6.19**   To connect the discharge wires, you have to either weld something that you can solder onto or solder directly onto the cell terminals.

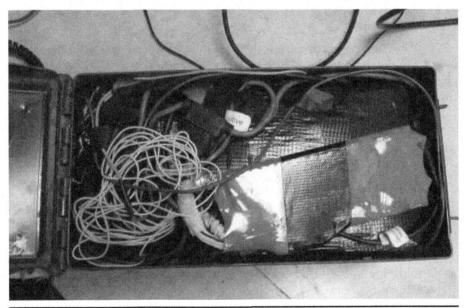

**FIGURE 6.20**   Here is the finish boost battery wedged in alongside the main pack using all available space in the battery box. The balance wires connect the two packs, and everything is controlled by the original BMS from the capacity pack. The BMS shunt can be beefed up with solder if the cutout is too low for the new battery.

FIGURE 6.21    The battery box with the lid shut. This was made from an Army surplus ammunition case.

When you heat-shrink the pack, the balance wires all will exit from the same place, and it will be impossible to know where they go just by looking at them. To determine the order in which to connect the wires onto the BMS, a multimeter should be used. Measure the voltage from each wire to the positive terminal, and use the voltage to tell where in the pack each wire is connected. In this way, there should be no unexpected sparks when connecting the balance wires.

### Pouch-Cell Battery Packs

Pouch cells may be soldered into a pack if they have long contact tabs. If the contact tab is too short, care must be taken not to melt the insulation around the tab where it enters the pouch; otherwise, there could be a short circuit. It's not possible to solder cells if the solder tabs have broken off. Use a high-power soldering iron and lead solder. Pouch-cell tabs are connected in the factory using ultrasonic welding, which requires a very specialized tool. If the tabs are cut short or broken, then there is very little you can do to repair the cell.

Figure 6.22 shows the insides of a pouch-cell pack. The numbered junctions are where the cell tabs are soldered together. In this case, the battery is 16 series, 3 parallel, so at each junction there are six tabs soldered together from different cells. The tabs of the cells are threaded through slots in small pieces of circuit board and then soldered together with a big blob of solder. The black foam you can see is padding between the cells. The cells actually look like the ones in Figure 6.22. To get the cells out, you usually have to cut the contact tabs because there are several cell tabs going through the same circuit board. It is possible to remove the cells without cutting the tabs if you can melt all the solder all the way through and pull the circuit board off all the cells at once. You have to solder suck and then cut up the circuit board joining the cell tabs, and bend it out of the way, to seperate the tabs. This is a difficult job and its easy to brake off cell tabs if you bend them when coated in solder. Usually you only want to unsolder cells from a pack if they are dead, so this is not a common problem. When building a pack out of pouch cells, group the cells into parallel cell groups first, and solder them together. Then assemble these in series to build up the pack. If you have fresh cells and fresh circuit board. The result should look like the pack in Figure 6.22. If you are reconfiguring and old pack then it wont look as pretty. Be very careful not to get the polarity wrong. If you do, you will be reminded by a spark and some smoke. Include balancing wires for the BMS. When the pack is complete, protect it with more layers of card, duct tape, or heat shrink and then padding, and put it in a strongbox for impact protection.

**FIGURE 6.22**    Pack made of pouch cells with tabs soldered together. The numbered junctions are where the cell tabs are soldered together. In this case, the battery is 16 series, 3 parallel, so at each junction there are six tabs soldered together from different cells. The black foam is padding between the cells. The cells actually look like the one in Figure 6.22.

### Soldering to Batteries

It is possible to solder onto metal cylindrical cells, but it is bad for the cells. This is so because it exposes the cells to high temperatures. Soldering onto metal cylindrical cells may be necessary if you have a dead cell in the middle of a pack or if the tab welding was of poor quality and has come undone. The problem is that the cylindrical cell conducts the heat away too quickly from the soldering iron, and the temperature doesn't go high enough to melt the solder. To solder onto batteries, a powerful soldering iron must be used along with good leaded solder that melts at a lower temperature than lead-free solder. This allows the soldering to be done quickly, which minimizes the damage to the cells.

If you are going to solder a battery pack, it is best to buy cells with solder tabs already welded to the cell, which makes the soldering easy. Soldering to batteries without solder tabs is the most difficult soldering job that you might attempt when building an electric bicycle.

Once the cells have solder tabs on them, soldering the tabs together to make a battery pack is easy. Follow the pack-building advice in Section 6.1.1, especially with regard to insulation and wire thickness. The battery tabs can be a point of failure in the pack if they are not connected in a secure way. Therefore, the sides of the battery where the battery tabs are connected need to be padded to protect against impacts, especially if the welding or soldering is poor.

### Battery Pack by Compression

Another way to build a battery pack is to use spring-loaded contacts to temporarily join the cells as in such commercial devices as cameras and radios. To scale this up for an electric bicycle, you can slide fat, high-capacity D cells down a 1.5-in-diameter PVC pipe. The cell connections are done using spring contacts sandwiched between the cells. To make the end pipe connections, you can use stoppers with a spring-loaded connection joined to an exiting wire that connects to other packs. The stoppers then can be duct-taped to the pipe, or you can use plumbing fittings to cap the pipe off. Make sure that the springs are thick enough to cope with the expected current and are made out of a low-resistivity material such as copper, nickel, or aluminum. This is quite a low-tech approach to battery packs.

### 6.1.2 Bypassing a Brushed Motor Controller

The simplest form of motor control is just a battery, a switch, and a brushed motor (Figure 6.23). Electric vehicles (EVs) like this have been around for hundreds of years. This form of control still has its uses in the age of microchips and cheap electronics. If your controller fails, and you need to rig

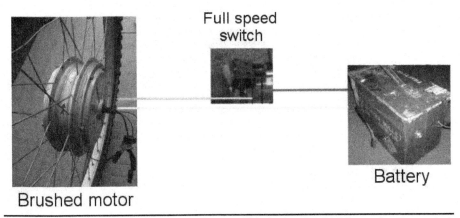

Full speed
switch

Battery

Brushed motor

**FIGURE 6.23**    Switch-controlled brushed motor. This is the simplest form of motor control.

something fast that will get you from point A to point B, there is nothing easier than just a switch. Some cheap electric bicycle motors are weak enough that you don't really need partial throttle anyway. The disadvantage is that you have to run thick power wires up to the handlebars and back. This is a crude form of cruise control. Using this form of control, you have to remember to switch it off when you want to brake because otherwise you might not be able to stop. If you want to go half speed then you have to cycle the switch on-off repeatedly. The switch will get spark eroded fast, so use the biggest switch you can buy, rated for the highest voltage and current you can find. Both sides of a double-pole, double-throw (DPDT) switch can be ganged together to double the current rating for this purpose. Use two switches in series to prevent losing control of the bicycle if the switch welds itself shut.

This simple control backup can be wired into the normal brushed controller with minimal extra wiring, as shown in Figure 6.24. On a brushed controller, one of the battery wires is always through-connected to one of the motor wires. Use a multimeter to continuity check and find which wire is through-connected to which motor wire. On my brushed controller, the red battery wire was through-connected to the yellow motor wire, but check on yours because it may be different. In this case, the controller can be by-passed with a handlebar-mounted switch connecting battery negative to the blue motor wire. This modification is also good for getting rid of unwanted pedal-assist mode.

### 6.1.3 Removing Pedal-Assist Mode

Pedal-assist (PAS) mode is a pain in the butt. PAS mode leaves you stranded on hills and makes the bicycle feel uncontrolled. Most PAS mechanisms use

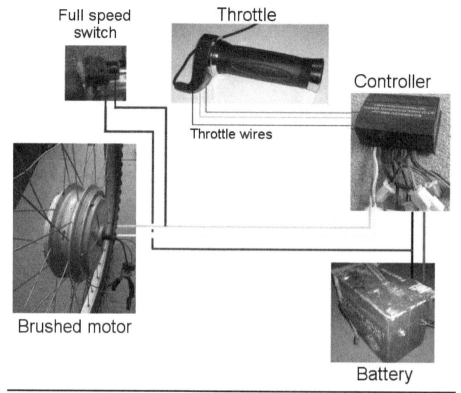

**Figure 6.24**  Bypass switch for brushed controller setup. This is a backup for the event of a controller failure.

a simple tachometer sensor to measure how fast you are pedaling, similar to the way a bicycle speedometer works. The sensor sends data to the controller, which also takes control inputs from brake-mounted electric stop switches and usually the hand throttle as well. It is sometimes possible to hack this sensor so that it thinks you are pedaling all the time or that you are pedaling faster than you actually are. This will turn the bicycle into throttle control only, which most people prefer.

If you have a brushed motor (where only two wires exit the hub), then it is possible to bypass the controller completely and control the motor using a switch. This could be instead of or as well as the controller and is a great way to bypass the PAS control (see Figure 6.23). Another way is to buy a new controller without PAS. This is easy and cheap with a brushed motor, and it's always good to have a spare controller because the cheap ones can be prone to failure. See the guide in Section 4.2.1 for wiring controllers. Sometimes there is an easy get-around on PAS-equipped bicycles whereby the tachometer can only tell if the pedals are rotating and not the direction of rotation. On these bicycles, you can free-wheel the pedals backwards, and this is enough to activate the PAS mode. Sometimes you can disable PAS mode just by unplugging the tachometer or crank sensor from the controller. The tachometer sensor is usually accessible just below the pedal crankshaft. On some bicycles, the crankshaft is enlarged, and the tachometer is inside it. This is a poor design because it makes it harder to access the tachometer and also makes the crankshaft components more specialized and difficult to replace when you hit the pedal on a trail rock and bend the crankshaft.

### Removing PAS Mode on Crank Rotation Sensor–Equipped Bicycles

The following instructions apply to the Schwinn I-Zip and other electric bicycles equipped with the crank rotation sensor. The sensor is connected to the motor controller with a three-wire plug that has a "black box" with a variable resistor on it. The three-wire plug's functions are:

- Black: Module power (+5 V)
- Red: Signal wire (+5 V)
- Blue: Signal wire (ground)

All that needs to be done is to permanently short the red and blue wires, and a signal will be generated indicating that the crank is rotating. The controllers used on this bicycle do not transmit information about crank speed—only that it is rotating. Then the bicycle can be used without the Pedelec. The bicycle will not function without the Pedelec module installed, however.

### 6.1.4 Overclocking/Upgrading a Brushed Electric Bicycle

So you have a 200-W Chinese electric bicycle, and it goes 15 mph. You are happy with it, but you feel that you could keep up with traffic better if it could go a little faster. Before you go and splurge on an all-new electric bicycle kit, you should have a go at overclocking the one you have to see if you can get any more speed out of it cheaply. There is always a risk of causing damage when pushing the performance envelope, but if the old bicycle kit gets broken in the process, who cares? You were going to buy a more expensive electric bicycle kit anyway! **Disclaimer:** Don't blame me if you follow these instructions and your stuff breaks. You do these experiments at your own risk.

To overclock an electric bicycle, you have to know what is limiting its power. Finding this out will involve a bit of testing and a process of elimination. There are several things that can limit power. These are listed below in order of probability for a commercial bicycle:

- Battery power limiting

- Controller current limiting
- Controller speed limiting
- Motor thermal tolerance limiting
- Motor rpms limiting

   The subsections that follow are a guide to how to overclock any brushed electric bicycle. The next section is a guide to overclocking brushless motor bicycles, but many of the tests apply to both. The tests are ordered with the easiest first and must be done in order for the process of elimination to be successful.

### Controller Speed Limiting

The first thing to check is the controller because this is the cheapest thing to replace. Despite European Union (EU) regulations and laws, most electric bicycles have no speed-sensing circuitry, and the controller does not limit speed. You can prove this by lifting the back wheel off the ground and doing a no-load test with a speedometer. You will find that the speed shoots up to higher than 17 mph. If the speed is limited to 15 mph, then the controller is limiting the speed. If so, great! Just buy a £20/$30 controller on eBay, and you can increase your bicycle's speed.

### Controller Current Limiting

To overclock an electric bicycle, you should buy a power meter (see Section 5.1.2) so that you can quantify what kind of power you have/want; otherwise, it's just a guessing game. With the power meter, you can determine whether the controller is limiting your power by wiring the battery straight to the motor via a handlebar switch (circuit shown in Figure 6.24) and comparing this power with that available when using the controller as wired normally. This test can only be done on a brushed motor, the ones with two wires that exit the motor. This controller bypass test probably will show no power difference, though, because the controller is not normally used to limit power on cheap commercial electric bicycles. If there is a difference, then great! Just buy a £20/$30 controller on eBay, and you will increase your power by whatever the difference was. In fact, it's probably a good idea to buy the high-powered eBay controller anyway because as you increase the power of the battery, you may find that the controller starts to become a limiting factor, or it might not take the higher voltage.

   There is usually no *governor* or bottleneck holding back a commercial electric bicycle because all its components are rated for around the same low-power output, about 200 W. The manufacturers do this to save on component costs. Having said this, the motor is not an expensive component, and most of them are rated fairly conservatively. The rated power for a motor may just be a number that was chosen to allow a factory to legally sell a motor in a specific market. The motor may have potential for much more power, maybe triple the power rating. The batteries are what usually limit the electric bicycle to a specific power rating. This is so because the batteries are the most expensive part, and the manufacturer does not want to waste money by exceeding their rated market specifications.

The batteries limit the power by their voltage. The voltage is also limited by the rated current of the batteries because the voltage will sag under load when you try to exceed the rated current. You can estimate the speed limit of your battery using the graph in Section 3.4.3 (Figure 3.16) if you know the volts and rated current. You probably will find this is close to the speed you can achieve under the conditions mentioned on the graph.

### Motor rpms Limiting

Once you have eliminated the possibility of the controller limiting your speed, you need to find out the battery voltage. There are two ways that battery voltage can limit your speed—either by hitting the maximum rpm limit for the motor (when going downhill) or by hitting a power limit versus wind resistance. A no-load test will tell you the maximum motor speed. Lift the wheel off the ground, and with a speedometer, measure the maximum speed. If this speed is close to the normal top speed of the bicycle, then the system is hitting the limits of the motor rpms for that battery voltage. This can be solved either by changing the gearing by relacing the motor into a larger wheel or by increasing the battery voltage. The no-load speed can be divided by the battery voltage to give a speed-per-volt specification for that motor. You then can use this to predict the outcome of increasing the wheel size or battery voltage. When you overcome the motor rpm limit, then speed next will be limited by the battery power limit versus wind resistance.

### Battery Voltage Limiting

If the no-load test speed was significantly higher than the normal top speed of the bicycle, then the bicycle's speed is limited by power versus wind resistance. This is probably the most common limiting factor for electric bicycles, and it means that you have to increase either battery voltage or battery peak current or both if you want to increase the speed. If the bicycle operates on a low-voltage battery (e.g., 24 V) and is still quite fast, then this means that there is lots of room for improvement simply by changing to higher-voltage batteries. The highest commercial electric bicycle voltage is usually 36 V, and the highest voltage you can increase it to is about 72 V. After that, it becomes impossible to find compatible controllers.

Doubling the voltage of the batteries will roughly double the power if the rated current output is the same. To increase the voltage, you just add more batteries of the same type and capacity in series. This is good news for

24-V, 250-W motors because you can double the battery voltage to 48 V, which will double the power, and controllers are still readily available for this voltage.

### Battery Current Limiting

With the power meter, see how much peak power you are putting through the hub motor during riding. Put the power meter on the handlebars where you can see it, and look to see if the voltage is sagging a lot when you accelerate. If you notice the voltage sag a great deal, then there is lots of potential for a power increase by buying more powerful batteries. Say that the voltage sags by 30 percent when accelerating hard. This means that if you buy higher-amperage batteries, you can gain up to 30 percent more power. Rated battery current is a product of discharge rate and rated capacity; for example, $3C \times 10 \text{ Ah} = 30 \text{ A}$. By increasing the battery capacity, which will increase the bicycle's range, you also can increase the bicycle's power. You can increase the rated battery current just by adding more batteries of the same type and voltage in parallel.

So you can increase the battery power in two ways—by increasing either or both voltage or rated current output. Say, for example, that your 24-V, 7-Ah, 1C NiMh battery (29 V fully charged) sags by 20 percent down to 24 V when accelerating under maximum load at 200 W. This means that if you buy a high-current 48-V battery, you could increase your power to 480 W. A high-current 60-V battery would increase your power to 600 W! And this is all with a 200-W-rated motor. With these higher voltages and currents, you will need a better controller. Sufficiently powerful controllers are available cheaply on eBay. To summarize:

$200 \text{ W} \times 48 \text{ V}/24 \text{ V} = 400 \text{ W}$ (this is the power increase from a voltage increase)

$400 \text{ W} \times 1.2 = 480 \text{ W}$ (this is the power increase from a current increase that eliminates the 20 percent voltage sag)

### Motor Thermal Tolerance Limiting

If you find after this that the motor gets hot (highly likely), then giving it more ventilation not only will stop it from melting but also will increase its power some more by reducing resistance in the windings and allowing more current to flow. To test this, you can touch the side of the motor after a ride and see how hot it is. If it's no hotter than the rest of the bicycle, then it can handle a bit more power. If you notice the motor getting hot, you also may notice a drop in power on the power meter as the windings heat up and prevent current flow. When overclocking the bicycle and pushing its limits, it's a good idea to buy an aquarium thermometer because this will give a more accurate temperature reading than just touch alone. Insert the probe into the hub where the wires go in. If the temperature of the outside of the motor is 40 to 50°C/104 to 122°F, then you can assume that the temperature inside is 70 to 80°C/158 to 176°F, near the melting point of some components, so don't exceed it. You can alleviate motor heating by drilling holes in the side covers to let the heat escape (Figure 6.25). Take the covers off first so that you don't drill through the motor windings and trash the motor. The windings are the part of the motor that heats up the most, so drill the vent holes nearest to them. With brushed hub motors, the windings are attached to the hub, which is good for heat dissipation. If you choose to do this modification, be careful not to weaken the side plates too much because they support the weight of the bicycle. Avoid using a motor modified in this way in wet conditions because it will allow water into the hub. Be careful with dust and metal debris, especially around your work area, where iron filings will be attracted into the motor by the magnets and could damage the bearings or cause increased friction.

Now you have reached the absolute limit to which you can take this 200-W motor.

This thermal limit may be around three to five times the continuous rating depending on heat dissipation. If you still want more power after all this, then you will have to buy another motor. The drawback of pushing a

**FIGURE 6.25**  Holes added to a small brushed motor to let the heat out and allow it to operate at higher power without melting.

motor to its limits in this way is that there will be a significant efficiency penalty. You may find that your range could be 10 to 20 percent longer with a larger motor. Figure 6.26 presents a flowchart summary.

### 6.1.5 Overclocking/Upgrading a Brushless Electric Bicycle

Overclocking a brushless electric bicycle is harder because brushless controllers are made specifically for brushless motors, and upgrading them can be difficult (see Section 4.2.1). It's not possible to bypass the controller like it is with a brushed motor to check for controller current limiting. However, we can still test for controller speed limiting, motor rpm limiting, battery voltage limiting, battery current limiting, and motor thermal tolerance limiting using the same processes as in the preceding section. Make sure that you have read the preceding section on brushed motors because most of the

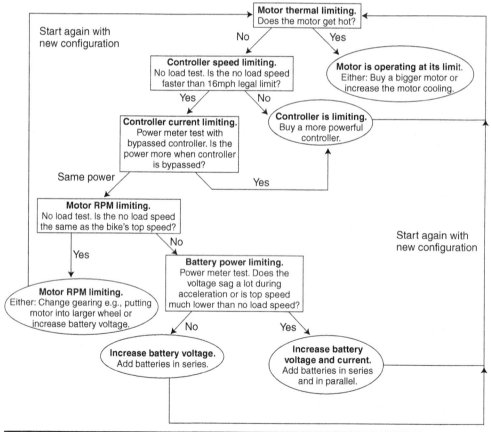

**FIGURE 6.26** A flowchart showing how to find what's limiting an electric bicycle's power and what to do to increase the speed and power.

tests apply equally to brushless motors, too. This will tell you what is currently limiting the power of your electric bicycle, which probably will be the battery power again. The problem is that if you increase the battery voltage, the controller may not withstand the higher voltage and may blow up.

### Test for Controller Current Limiting on a Brushless Motor

First, do the tests to eliminate the possibility of controller speed limiting or motor rpm limiting. Then, if under full load there is no significant voltage drop and the motor does not get hot, the controller is probably limiting the current. If the peak current is always the same no matter what state of charge the batteries are in, then you can be fairly certain that the displayed current on the power meter is the current limit of the controller.

If the controller is not potted in resin, then it may be possible to lift the ampere limit on it by soldering a shunt (see Section 6.1.6). If you solder a shunt, you also might want to install a variable-current-limit pot for more control (see Section 6.1.7). You can use a power meter in conjunction with a variable-current module to accurately set the current limit to whatever you think the controller and motor can withstand. Be sure to check that the field-effect transistors (FETs) can handle this current by checking their specifications. Make sure that you have adequate cooling on the controller, and monitor the heat buildup in the controller. If heat is a problem, you can attach heat sinks from an old computer or computer power supply.

If you notice a large voltage sag when under load, then you can safely increase the power by increasing the rated current of the battery (see the example under "Battery Current Limiting" in the preceding section). You can increase the rated current of the battery just by adding more batteries of the same type and voltage in parallel, this also will increase the bicycle's range, or by swapping for a higher-amperage battery.

### Increasing the Voltage

If the controller is not potted in resin, then it's possible to open it and look at the voltage ratings of the capacitors to get a good idea of how much of a voltage increase the controller can take. The upper voltage limit of the controller is usually around 75 percent of the capacitor value. You also can determine the model number of the FETs and find their specifications, which will tell you about the voltage of the controller. The number on the FETs will be something like IRFB4110, and you can look this up in the Table 6.1 or on the Internet to find the voltage rating. The upper voltage of the controller will be something like 75 percent of the voltage limit of the FETs. If the controller is potted in resin, you can usually soften it up with paint stripper over a few days, and then scrape it off with a knife. Be careful and use the proper PPE (personal protective equipment). If you find that the controller has high-voltage-rated FETs and capacitors, then great! You probably can increase the voltage and receive a power boost proportional to the voltage increase. If the FETs are not rated for high voltage, then you have two options: You can either upgrade the components of the controller or you can buy a more powerful controller and try to make it work with your motor. Alternatively, at this stage, if you are working with a cheap Chinese electric bicycle setup, you may want to consider just selling the bicycle and starting fresh with a new high-power electric bicycle kit.

### Upgrading Controller Components

Once you find your speed restricted by voltage, either by motor rpms or by battery power, you will want to increase the voltage. However, with a brushless motor, you may be limited by how far you can safely increase the voltage without causing damage to the controller. In this situation, you could consider moding (modifying) the controller to increase the voltage limits of the components. This is usually attempted only by those "crazy people" who really want to stretch the limits of electric bicycle technology. These people already have the expensive high-power motor and controller, but they want to make them even faster—at any cost! This will involve swapping the FETs for higher-voltage ones and/or upgrading the capacitors. The how-to guides for these changes are shown in Section 5.4.1 for FETs and Section 5.4.3 for capacitors. The upgrade procedure is the same as repairing the controller from a blowout. Doing these modifications may allow you to increase the voltage the controller can take, but there is no guarantee that it will not blow up.

### Heat Dissipation in Brushless Hub Motors

The problem with heat dissipation in brushless hub motors is that the windings are in the middle, and there is an air gap between the windings and the outside of the motor. Air is a good insulator and causes a problem for heat conduction. You can drill holes in the side covers of the motor, as mentioned earlier. This is even more effective for cooling brushless motors than it is for cooling brushed motors. The other option with direct-drive hub motors is to fill them with oil. Filling the motor with oil will increase the conductivity and allow the heat to conduct out through the outer surface of the motor, which should increase heat dissipation dramatically and cool the motor. You don't have to completely fill the motor with oil. Just a quarter full will be enough because centripetal force will make the oil flow to the right places.

## 6.1.6  Increasing the Power Limit by Soldering the Shunt Resistor

This is a simple modification that tricks the controller into thinking that the current is lower than it actually is so that you can get more power from it.

Adding extra solder to the shunt reduces its resistance and thus lowers the voltage drop that the controller measures. However, this is a crude modification that could damage the controller if too much current is forced through its components. This modification can be made a bit more refined if a parallel wire is used, and a switch is added that can activate the parallel wire, or one could shortcut half the shunt length with a wire and an on/off switch. Similarly, you can reduce the current limit by cutting some of the wires in the shunt or filing them to reduce their thickness.

1. Locate the shunt resistor. It will look something like the four thick wires standing out from the board shown in Figure 6.27. There may be more or fewer wires on different controllers, but they always stick out from the board and should be easy to spot.
2. Simply heat up the shunt with a soldering iron, and add solder to beef up the thickness. The current limit generally is proportional to the cross-sectional area of the shunt.

### 6.1.7 Variable-Current-Limit Modification

All controllers can have their current limit changed with alterations to the shunt. This is a very crude way of doing it because the resistance of the shunt is very low, and it has lots of current flowing through it, so there are not

**Figure 6.27** The shunt resistor can be found easily on all controllers. It looks the same—like thick parallel wires. On the V1 Crystalyte controller, there are four wires; on the V2 Crystalyte, there are two wires.

many things that we can do to it. Most controllers can be modified to change their current limit by changing other resistor values as well. The controller always compares the voltage drop over the shunt to the voltage drop across another resistor. If we locate this resistor, we can change its value, using a variable resistor, and this will give us a knob that we can twist during a ride to make the bicycle go faster or slower.

### Variable-Current Crystalyte V1 Controller Modification

This modification can be used to vary the current limit in the downward direction. This modification is specific to the Crystalyte V1 controller but should work on any controller using the KA3525 comparator for current limiting. This modification works well with the shunt soldering modification that alters the current limit in the upward direction. With these two modifications, you can set whatever current limit you feel like. The shunt modification will increase the upper limit for better top speed, and then you can dial it down for everyday use by using this modification. Thanks to super electrical engineer Richard Fechter from San Francisco for discovering this modification.

1.  Locate the KA3525 comparator chip on the controller circuit board. Find legs 1 and 16 on the chip. Leg 1 is marked on the chip with a dot; leg 16 is opposite leg 1, as seen in Figure 6.28.
2.  Locate the battery negative on the controller, as shown by the ground in Figure 6.28.
3.  Solder wires onto pins 1 and 16 and the battery negative.
4.  Get a 150-k$\Omega$ resistor, a 51-k$\Omega$ resistor, and a 10-k$\Omega$ potentiometer.
5.  Wire the circuit as shown in Figure 6.29, it should look something like what is shown in Figure 6.30. Pin 1 on the KA3525 chip goes to the middle pin on the potentiometer, the one that can change resistance relative to the other two legs. Alternatively, you could use a switch and a fixed resistor to provide a boost switch if you want a solution that is easier to operate during a ride. This could be used to switch between a legal power limit for road use and a higher power limit for off-road use.
6.  Locate a hole in the case of the controller. In Figure 6.31, for the V1 Crystalyte controller, the hole for the reverse switch was used because the reverse function is never used. Alternatively, you would need to drill another hole in the case. Remember that there needs to be room behind the hole for the potentiometer when the circuit board is back inside the case. Don't drill the case with the circuit board inside because you could break it.
7.  Put everything back together, and mount the variable resistor. Secure it in place with its lock nut.

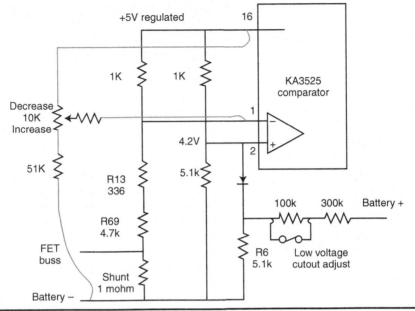

**FIGURE 6.28** Where to tap the lead for a variable current limit for the Crystalyte V1 controller.

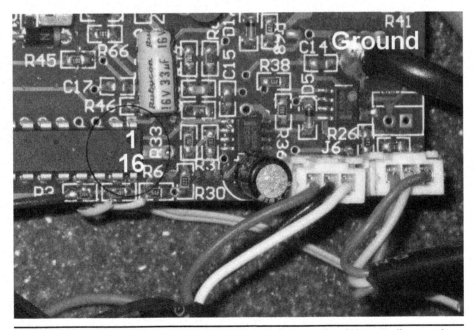

**FIGURE 6.29** Variable-current-limit modification on a Crystalyte V1 controller or others with the KA3525 chip.

Chapter Six

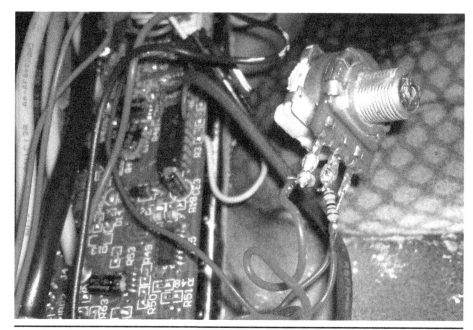

**FIGURE 6.30** This shows the KA3525 chip with pins 1 and 16 taped for the variable-current-limit modification. The 10-k$\Omega$ potentiometer is shown in the foreground.

**FIGURE 6.31** This is how the potentiometer is fitted into the case where the reverse button was located.

## Crystalyte V2 Controller

On the V2 controller, the comparator is controlled by a different chip. Resistor R4 is compared with the shunt. By changing the value of R4, you can alter the current limit of the controller.

1. Locate resistor R4 (Figure 6.32).
2. Unsolder R4.
3. Replace R4 with a variable resistor. R4's original value was 1.2 kΩ. To increase the current limit, you can increase the value of R4. To decrease the current limit, you can decrease the value of R4. Putting a 1-kΩ variable resistor in series with a 500-Ω resistor at R4 would allow you to adjust the limit downward. Putting a 1-kΩ variable resistor in series with a 1.2-kΩ resistor at R4 would allow you to adjust the limit upward. Putting a 5-kΩ variable resistor in series with a 200-Ω resistor at R4 would allow you to adjust the limit up and down. Be careful how far you twist the throttle with a 5-kΩ potentiometer, though, because it could allow the controller to be taken outside its safe working limit.
4. Run long wires from the variable resistor to the solder pads at R4.
5. Drill a hole in the case to mount the variable resistor. Remember that there needs to be room behind the hole for the potentiometer when the circuit board is back inside the case. Don't drill the case with the circuit board inside because you could break it.
6. Put everything back together, and mount the variable resistor. Secure it in place with its lock nut.

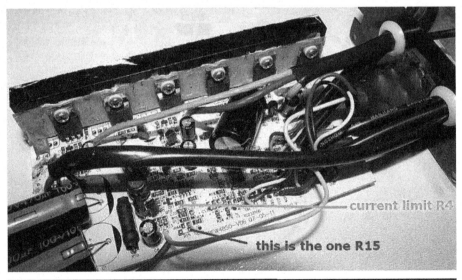

**FIGURE 6.32** Position of R4 on the Crystalyte V2 controller. Install a 1-kΩ potentiometer here to add variable current limiting.

### Shenzhen E-Crazyman Controller

To make the current limit adjustable, it appears that changing the value of R8 should do the trick.

1. Locate R8 on the board (Figure 6.33), and remove it. Attach a pair of small wires to the spot where it used to be. **Note:** The position for C7 above it is parallel to R8, so the wires can go across both spots.
2. Bring the other end of this pair of wires out to a 50-k$\Omega$ potentiometer that has a 15-k$\Omega$ fixed resistor in series with it. The original value of R8 was 18 k$\Omega$, so you really should use an 18-k$\Omega$ fixed resistor to keep the same maximum. A 15-k$\Omega$ resistor is a more common value and will give you a slight increase in maximum current. The value of the potentiometer will determine the adjustment range. If you used a 20-k$\Omega$ potentiometer instead, it would allow the adjustment to go down to about half the maximum. A 50-k$\Omega$ potentiometer should take it way down.

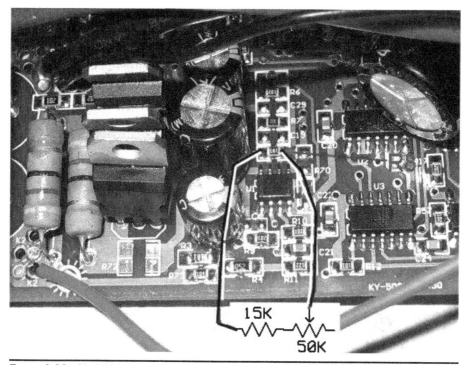

**Figure 6.33**  Variable current limit for the Shenzhen E-Crazyman controller.

## 6.1.8 Modification of a Battery Charger

It is possible to prevent sparks during charger connection by fitting an in-rush thermistor. An in-rush thermistor is a component that slows down the fast in-rush of current when a device is switched on by heating up and quickly increasing its resistance. Normal current is allowed to flow because the thermistor cools down when the capacitor voltage has built up and the current slows to a normal level. A thermistor can be fitted either into the charger wires of the battery or on the power input side. Choose whichever side causes the sparking problem. Make sure that the thermistor is rated for the application. This will improve the life expectancy of some of the poor-quality battery chargers.

Improving cooling is another modification that will increase the longevity of a charger. Monitor the temperature of the charger, and if it feels hot on the outside, then it is likely to be even hotter inside. Amazingly, some chargers are built in sealed boxes with no airflow, and they get very hot. You can cool a charger down by taking it apart and cutting holes for passive cooling or mounting a fan. The power for the fan(s) can be taken from the battery side if several fans are series-connected to match the battery voltage.

### 6.1.9 Do-It-Yourself (DIY) Battery Chargers

Building a DIY battery charger is possible if you know what you are doing, but it is not recommended because of the risk of death by electric shock. Commercial electric bicycle chargers are very cheap and recently have improved in quality, so it is not necessary to build DIY chargers any more. However, a few years ago, some brands of charger were very unreliable and seemed to break every 6 months, so it was worth experimenting then. It's also possible to build faster, more powerful battery chargers than those typically available commercially. Typical electric bicycle battery voltages are generally safe. It's difficult to get a shock from 72 V because dry, intact skin is a good insulator. Main circuit voltages are much more dangerous. Do not work on anything that is connected to the main circuit; always disconnect it first! Attempt to build a DIY battery charger only if you know what you are doing with electrical stuff. Always use a circuit breaker and an earth connection. A fuse will not prevent you from getting an electric shock. I cannot be held responsible for any accident or injury caused by your work.

### How to Build Your Own Battery Charger

A battery charger can be made from a normal power supply if the voltage is right for the battery. The voltage of the charger should be exactly that of the constant-voltage part of the charge curve. There also should be a method of limiting the current for the constant-current part of the charge curve so as not to exceed the current limits of the battery or power supply. One method to limit current is to use resistance such as a light bulb. This can be very inefficient, dissipating lots of energy as heat, and it does no favors for the green image of electric bicycling. The efficiency will depend greatly on the voltage wasted across the light bulb and its ratio to the useful energy used charging the battery.

In fact, the simplest battery charger is simply a diode and a light bulb in series with your battery from the main circuit supply. However, this is very dangerous because it has no isolation transformer. In popular electronics, anything like this is called a *widow maker* because it gives all terminals of the battery a live connection to main circuit voltage, which will electrocute anyone who touches it. Do not do this! It's also very wasteful—something like 90 percent of the energy is wasted as heat.

To make a good battery charger, you should use a power supply that is around 20 V higher than your battery voltage, and then use a light bulb to slow the current to make it comfortable for your battery. The 20-V difference gives the battery room in which to charge up. As the battery's state of charge increases, its voltage will increase and take up more of this room. As this occurs, the bulb will grow dimmer. Higher-voltage batteries will need more room to grow because they have more cells. A 48-V NiMh battery can be anywhere from 38 to 60 V depending on state of charge. If the battery is heavily discharged, it may put too much voltage across the bulb and cause it to blow. If you find this to be the case, use two bulbs in series.

The power supply can be either a main circuit transformer or a switched power supply, but it needs to be able to handle the amperes at which you want to charge. Main circuit transformers will need a bridge rectifier to change the power over to direct current (dc). For limiting the current, low-voltage MR16 bulbs work well. They are just right to dissipate about 15 V and limit current to a few amperes depending on the power of the bulb. A 10-W bulb will slow current to 1 ampere, a 20-W bulb to 2 amperes, and a 50-W bulb to 5 amperes. To make it safer, you should attach a trip-switch circuit breaker to the plug to prevent electrocution if anything goes wrong.

*Making the DIY Battery Charger*
1. Chosoe a power supply with a voltage just above that of your fully charged battery. For a 48-V battery, you would want a 60-V supply; for a 36-V battery, you would want a 48-V supply; for a 24-V battery, you would want a 30-V supply; and for a 72-V battery, you would want a 110-V supply. You can use a switched power supply or, if you have a 72-V battery, you could use a main circuit transformer to get 110 V. If your main circuit is already 110 V, you should use an isolation transformer.
2. Wire up the plug if the power supply doesn't have one.
3. If you are using a main circuit transformer, the alternating current (ac) must be converted into dc using a bridge rectifier circuit. Bridge rectifiers can be bought cheaply from a component supply store.

4. From the dc output, connect the light bulb of the desired power rating. If the bulb is above 10 W, then it will need to be mounted properly because it will get hot enough to melt plastic or burn table tops, carpet, etc.

5. Next, connect a charging plug that will fit your battery. Charging plugs can be bought from a component supply store.

6. Decide on the charge-termination method depending on what battery chemistry you have to charge.

7. For charging NiMh/NiCd batteries, a thermostat circuit needs to be connected to the power output. Cheap thermistor-based thermostat circuits can be bought from a component supply store. Set the temperature cutoff by putting the thermistor into a warm water bath at 40°C/104°F and then changing the variable resistor with a small screwdriver until you find the sensitive zone where you hear the relay click on and off with minor changes of the potentiometer. Connect the relay to control the power output from the power supply. The thermostat circuit will need a power supply of its own because it usually works on 12 V. This supply should be powered by the main circuit so that it is always functioning.

8. For charging most other battery chemistries, a constant-voltage termination can be set using the adjustment potentiometer on the power supply. Find the voltage potentiometer, and adjust it with a small screwdriver, monitoring the output with a multimeter. The power supply should be set for exactly the voltage of the fully charged battery and no more.

9. Assemble and mount it all inside a nice case. The ugly attempt in Figure 6.34 should not be used as an example.

**Figure 6.34** My experiment with a DIY battery charger. The power of the light bulb determines the charge rate. Charge termination is provided by either voltage match of the supply to the fully charged battery or a thermostat circuit.

### Balancing Cell Level Charger

Mobile phone chargers are designed to charge individual lithium cells. A good battery charger for a lithium battery can be made from multiple phone chargers connected to each cell group that makes up the pack. This will balance the pack by charging the cells individually and charge each cell to the same constant voltage. The cell chargers can be wired to a single multipin plug for ease of connection. You just need to make sure that the battery chemistry, and therefore, the voltage, is the same as the type for which the charger was designed. Power supply outputs can be connected in series with each other only if they are electrically isolated. *Isolated* means that the output is isolated from the main circuit. If you connect outputs from nonisolated power supplies in series, then a short circuit of the main circuit supply occurs. This will blow a fuse or trigger a circuit breaker, and your family/housemates will not be pleased. To test that a power supply is isolated, you can continuity test with a multimeter from the main circuit pins to the dc output. If there is any connection, then the power supply is not isolated.

## 6.1.10  Racing Electric Bicycles

### Build Tactics

Racing electric bicycles is a much heavier duty than normal commuting to work. When racing, things such as tires and brakes become very important to your lap time. The best thing to do is to build your bicycle specifically for the racetrack. If there are tight turns on concrete roads, then slick tires and disc brakes are essential. If the race is uphill, then low gearing is essential. If the race is off-road, then knobby tires are important. If it's a distance race, then high gearing is essential. Some races allow you to recharge halfway through, so fast charging may be essential.

If you know the route beforehand, then you can use the speed and distance theories mentioned in Sections 3.4.3 and 4.1.2 to design your ride specifically for the racetrack. For example, you could optimize the battery capacity versus weight tradeoff to only have enough energy for the race and carry no unused battery weight. You could calculate the optimal speed that is achievable for a given weight limit. If the race is short, then it will be decided by the best battery rating and braking/cornering. If the race is long or there is a strict weight limit (e.g., 100 miles or 40 kg/88 lb), then pedaling and personal fitness will be important. Reliability is always important in racing because as you push the bicycle to its limits, faults start to show that you may never have noticed before.

Aerodynamics are always important in racing but are very difficult to build into an existing bicycle. The best aerodynamics tactic with electric bicycles is to use a bicycle that allows you to stay close to the ground. Cyclists do this using recumbent bicycles, where they are seated between the wheels, because this position is best for pedaling. When racing high-powered electric bicycles, there is little point in pedaling, so it's possible to use BMX (bicycle motor cross) or folding bicycle frames and adopt a minimoto riding position for improved aerodynamics. Racing bicycles with their 700C wheels actually have very poor aerodynamics by comparison. See the performance aerodynamics modifications in Section 6.1.12.

Safety gear also should be a priority in a race because you are likely to take more risks than usual, such as trying to outbrake or outcorner an opponent, and there is a higher chance of a collision or of falling off. It is suggested that racers wear heavy trousers (jeans or leathers), leather gloves, and elbow and knee pads, as well as the usual helmet.

### Racing Tactics

There are three ways to beat an opponent in a race: outcorner, outbrake, and outrun. Take the racing line through a corner. Go wide to begin with, and then turn into the bend as close as possible to the apex. Then go wide when coming out of the turn. This will allow you to cut the corner and turn the least sharply, thus permitting you to corner at higher speed. When cornering, you should seek to outbrake your opponent by continuing fast and leaving braking until the last possible moment while still being able to make the turn. When coming out of a turn, you should go back to maximum power straightaway in order to get up to speed as fast as possible.

On a tarmac surface, slick tires will help with cornering and allow you to outcorner your opponents. Knobby tires will buckle when leaning and will lose traction, causing the wheel to slip and the rider to fall off. Higher tire pressures are recommended for road racing and lower pressure for dirt-track racing. Use at least tire sidewall pressure for road racing and 10 psi less than rated for dirt-track racing. A well-adjusted set of front and back disc brakes will give you the best stopping power to outbrake your opponents. Choose the largest rotors available to dissipate the heat and provide more braking surface.

Obviously, power is the most important factor, so build the bicycle with the maximum power that the rules will allow. Weight limits are more common because they are enforceable, unlike power limits, which are impossible to discern. Power-to-weight ratio is of secondary importance to absolute power. You should aim to use all your charge in the race so that you don't carry unnecessary weight or go slower than you needed to. You should aim to use 80 percent of your charge and have 20 percent in reserve so that you don't run out and look silly. Regenerative braking is a really good idea to have when racing electric bicycles because it probably will save 20 to 40 percent of your power depending on the amount of cornering in the course. Regenerative braking will give you a much better energy-to-weight ratio over other bicycles and increase your braking, allowing you to outbrake your opponents, too.

*Prerace Checks*
- Check that your batteries are fully charged.
- Check that your tires are inflated to the optimal value.
- Check that you have the correct tires for your racetrack.
- Check that your brake pads are not worn.
- Check that your brakes are adjusted correctly.
- Check that your fuses are rated high enough to withstand the extra current used in racing. You don't want to have to swap fuses during a race.
- Check that all the components in the path of the main battery current are capable of handling the current that will be pushed through them.
- Check that the BMS doesn't cut out inside the current limit of the controller. You don't want to have to unplug/replug your battery in a race or be worried about using the full power of the controller.
- Check that everything is securely fixed to the bicycle, including batteries, quick-release wheels, seat, etc.
- Check that the folding mechanism is tightened securely.
- Check that the controller has sufficient cooling so that it doesn't overheat.
- Check that the batteries don't overheat and are within (or close to) their operating specifications. Decide beforehand how much you are going to abuse them before the race. Remote control (RC) racers get a much shorter battery life because of the abuse they heap on their batteries during a race.
- Check that the motor doesn't overheat and is within (or close to) its specifications.
- Check that you have brought all the necessary backup tools and parts in case something goes wrong before or during the race.

## 6.1.11 Extending Range: Hybrid Electric Bicycles

It's possible to overcome the range limitations of batteries if you have a gas generator set strapped to the back of your bicycle. The advantage is that you can fill up at any gas station on your way and continue to ride for days at a time. The disadvantage is that the bicycle no longer feels like a bicycle when it is making loud throbbing, revving noises. You end up riding something that is not as cool as a motorbike or a bicycle. There are also legal implications to putting a motor on a bicycle that probably make it illegal in most countries and states without proper tax and registration. With current lithium batteries, we can comfortably build bicycles with a range of 50 or even 100 miles if we really load the bicycle down with batteries. A big enough generator basically can give the bicycle infinite range. However, on a bicycle frame, the maximum safe riding speed is 25 to 30 mph, so a 100 mile trip is going to take more than 3 hours. Who wants to ride for 3 hours?

A much better option is to take the train and bring your bicycle with you for the short journey at the other end. Anything containing gas would not be allowed on public transport because of the fire risk and fumes. Gas generators are usually bulky and heavy. A large power convertor is also needed to change the voltage into something usable for the batteries. The space taken up by a generator and voltage convertor probably would be equal to an extra 50 miles worth of battery capacity; therefore, it's usually much easier to just add another battery pack. The only application suited to a gas range extender would be for places where there is no public transport available or for long trips into the wilderness. Here, though, a full-on motorbike might be more appropriate.

A different application for a low-powered generator would be remote charging. This might be useful for camping in the wilderness, where there is no local electricity supply. It's perfectly okay to charge and discharge a battery at the same time. Performing the function of power buffer between a load and a source is an easy task for any electric bicycle battery, so only a small one is needed.

There are several ways of using a generator as a range extender. To provide the main power to extend vehicle range or for opportunity charging without significant range extension, power conversion is needed because generators provide main circuit voltage, which is usually to high to directly charge an electric bicycle battery. Some generators also provide a 12-V output for car battery charging, but this is too low to directly charge most electric bicycle batteries, and the power output is too low as well. One way around this voltage-conversion issue would be to use a 110-V system and a bridge rectifier for ac-dc conversion. However, it is difficult to find commercial electric bicycle controllers that are capable of handling this voltage. Some generators are overrated and can only operate for a set time before they overheat.

### How to Build a Gas Electric Bicycle Range Extender
1. Find a generator with a power output near to peak power demand.
2. Find a large charger or power supply for use as a voltage converter that is equal to or just under the power rating of the generator. Don't exceed the generator's power rating because it will overload the generator. The output voltage should be right for charging the type of battery you have. The input voltage should match the generator's output. This setup will require bigger or multiple chargers compared with normal charging requirements.
3. Connect the battery, charger, and controller together in parallel using a nice junction box. Avoid using small-gauge wires designed to carry a normal battery-charging current because they may be too small.

4. Plug in the charger(s) to the generator.

5. Fix the generator and charger to the bicycle. A rear rack or a separate trailer are probably the only places the generator will fit. The fixture could be either temporary or permanent.

## 6.1.12  Improving Aerodynamics on an Electric Bicycle

Because a bicycle is so light, the main factor that wastes energy is not stop/start acceleration as in a car. Wind resistance is the main cause of energy loss for bicycles. Cars have a smooth body shell that reduces drag, whereas a bicycle with a rider is a horrible shape for moving through air. The drag coefficient of a bicycle is about twice that of a car, and this causes bicycles to drag lots of air behind in their wake. The aerodynamic shape of the vehicle matters more when it is traveling at high speeds. Having a more aerodynamic shape will allow a vehicle to use less power or travel faster on the same power.

Aerodynamics are important in car design, which is why all the green vehicles look funny. And aerodynamics are a major factor in competitive cycling, which is why cyclists do silly things, such as shaving their legs, wearing skintight Lycra, wearing silly shaped helmets, and riding silly shaped bicycles. Fortunately, we electric bicyclers have a motor for speed, so we don't have to use these silly tactics to increase our speed. However, if you do want to turn to the dark side, then here are a few tips for racing.

The equation for the force of drag acting on a moving vehicle is given as Equation 6.4. This shows that the area and drag coefficient are both equally important to the drag force on the vehicle. We can reduce either or both area and drag coefficient to increase the speed or efficiency of our vehicles. This equation is just for illustrative value. No one expects you to use it. The maths determining speed and power were made simple in Section 4.1.2 and in the graph in Figure 4.3.

---

**EQUATION 6.4**   Force of drag on a vehicle.  Where $\rho$ = air density, u = velocity, $C_d$ = drag coefficient, A = frontal area.

$$F_d = \tfrac{1}{2}\,\rho\,u^2\,C_d\,A$$

### Reducing Frontal Area

Reducing frontal area is the easiest action to take because it's easy to measure. Here are some tips:

- Use a recumbent bicycle, where the rider sits in between the front and rear wheels. Avoid tricycles, which have a much larger frontal area.

- Use a low-riding bicycle frame with small wheels, such as a BMX, kid's bicycle, or folding bicycle. Racing bicycles aren't needed if you're not going to pedal. The height of the racing bicycle allows the rider to pedal.
- Reduce the saddle and handlebar height to minimum.
- Fit smaller wheels to get closer to the ground.
- Take the seatpost off, and sit on the rear rack if it is strong enough.
- Use dropdown racing handlebars, and assume a tucked riding position.
- Avoid large pannier bags that extend out sideways into the airflow.
- Avoid mounting batteries where they force your legs further out into the airflow when pedaling.
- Try to mount the battery and controller either behind the rider on the rear rack or in front of the rider in a basket.
- Avoid mirrors or reflectors that extend out into the airflow.
- Avoid baggy clothing such as open jackets that collect air like a sail or baggy trousers.

### Reducing Drag Coefficient

Reducing drag coefficient is difficult to quantify or visualize. The ideal aerodynamic shape is a teardrop with the tail pointed behind and the round surface in front. The way it works, without getting too technical, is that it allows the air that was forced around the teardrop shape at the front to recombine more gradually at the tail. The elongated tail of the teardrop shape allows time for the separated air to gently recombine without getting all mixed up. If the transition at the rear of the vehicle is too steep, then the airflow will not pass round the vehicle smoothly, and the vehicle will drag loads of air along behind in its wake.

It's the rear of the object that causes most of its drag; the front is less important. The Toyota Prius and other new fuel-efficient cars use an angle of about 12 degrees from the top of the roof to the trunk. This seems to be the magic angle, above which turbulence occurs. Obviously, it would be impractical to have a massive long tail all at 12 degrees until the top and sides meet up, so car designers compromise by chopping some of the tail off. The shape doesn't have to be perfect to still give some benefit.

It's possible to reduce drag by shaping individual components to be smoother and teardrop-shaped. However, covering the whole structure with one big teardrop shape, called a *fairing*, yields much better results. Some of the things that reduce frontal area also may reduce drag coefficient as well, which would multiply the effect on drag force. Therefore, some changes can have a big effect. For example, taking a wing mirror off a car gives a fuel-efficiency bonus of about 7 percent. Keep in mind that not all parts of the bicycle are traveling at the same speed. The wheels and spokes spin around much faster and are thus subjected to different airflows. To reduce drag coefficient, you can:

- Build a large teardrop-shaped fairing around the whole vehicle (see Figure 6.35).
- Build a teardrop-shaped tail fairing.
- Build a front fairing.
- Wear a rucksack with a teardrop tail shape.
- Buy a teardrop-shaped helmet.
- Fit wheel coverings or "aero" wheels (marginal effect).
- Use thin tires (marginal effect).
- Wear skintight Lycra (marginal effect).
- Put Cling-Film or heat shrink around rough edges such as bolts, etc. (marginal effect).

Using all these aerodynamic tricks, it's amazing what can be achieved. The Shell Mileage Marathon is a competition for low-carbon vehicles to use as little energy as possible around a racetrack with a minimum speed limit. There is an electric bicycle class in which some of the vehicles can reach efficiencies of 1.5 Wh/mile. The world record for a human-powered vehicle is held by Canadian cyclist Sam Whittingham. With no motor, just pedaling on a flat road, he set the record at 81 mph using a special low-rider recumbent bicycle with full fairing (Figure 6.35). Figure 6.36 shows an unusual electric bicycle designed for drag racing.

**FIGURE 6.35**  With good aerodynamics, even a low-power electric bicycle can get you very high speeds. This human-powered vehicle (no motor) reached up to 81 mph on flat ground to set the world record. The rider is Sam Whittingham with his bicycle, the *Varna Diablo*.

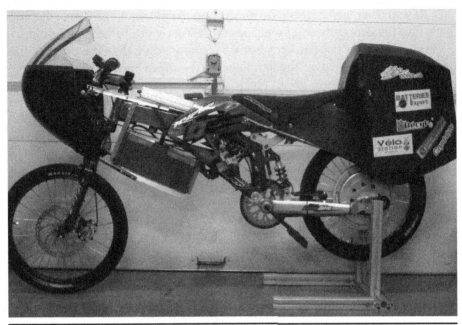

**FIGURE 6.36**   Doctor Bass's Killabicycle with front and rear fairings. This electric bicycle was designed for drag racing.

## 6.2  Safety Modifications

### 6.2.1  Regenerative Braking and Plug Braking

*Regenerative braking* (regen) uses the motion of the vehicle to recharge the battery, and as a consequence, this slows the vehicle. This is one of the main things that makes hybrid cars more efficient. Some of the expensive electric bicycle controllers have regenerative braking included as part of the controller, and other controllers can be modified easily to have regen. Regen braking is simplest when it is done through the controller, but it also can be done by bypassing the controller. Plug braking is similar to regen braking, but the kinetic energy is wasted across a resistor instead of charging the battery.

Regen and plug braking are useful for electric bicycles if you want extra stopping power or if your mechanical brakes are maintenance-intensive. For safety sake, don't rely on any one braking system; have backups. Before you install regen or plug braking on your ride, you must ensure that your bicycle has rock-solid dropouts that can take the punishment. The torque leveraged on the dropouts during regen braking is colossal, and anything other than thick steel dropouts with added torque arms is going to get damaged.

If you use regen braking on aluminum dropouts, it will cause the axle to cut into the dropouts and spin. This will destroy your frame and your motor. If you have this happen on a front-mounted motor, then it will break the forks, and you will go over the handlebars. Check out Section 6.2.2 on torque arms.

Regen braking on an electric bicycle is sometimes a bit of a gimmick. You will gain only a small percentage of the total energy with regen braking, maybe 5 to 10 percent of the energy used in normal riding. This is so because electric bicycles use most of their energy fighting wind resistance, not stopping and starting, and they weigh very little. A car or a bus weighs a lot more and usually stops a lot more, so regen braking is more important there.

### Implementation in Controllers

Some electric vehicle (EV) controllers have regen braking selected all the time. This means that as soon as you release the throttle, the car will start to slow down as if it were a gas engine car with engine braking. The logic behind this is that the EV has no engine braking and feels slower than a gas-powered car because the driver feels less difference between throttle pedal on and throttle pedal off. You can only "feel" accelerations/decelerations, not constant speeds, and the driver misinterprets this. This perception of reduced power is not good for car sales, so manufacturers install controller settings that mimic engine braking by using regen braking. I think this is a bad thing because coasting makes electric cars more efficient, which should be their main selling point. In fact, some efficient drivers of gas-powered vehicles use engine-off coasting to save fuel at times when the engine is not needed. On a gas-powered vehicle, it is necessary to turn the engine off while coasting to save fuel, which can be dangerous. Leaving the engine at idle still uses lots of fuel. Some car manufacturers have developed mild hybrid systems that automatically switch the engine off when it is not needed to save fuel. Unnecessary regen braking is a waste of energy because of the conversion inefficiency of transferring the energy back into the batteries. Hopefully, in the future, mimic engine braking will be an optional feature, not a default.

### Making a Plug Braking System

If your controller does not support regen, then you can use this modification to add a plug braking switch for more stopping power. The modification can be done on either brushed or brushless motors, but there is one extra step involved for brushless motors. The idea is to use a relay to short out the coils of the motor across a resistor. The severity of the braking is determined by the resistance. The most severe braking can be achieved without a resistor. Without a resistor, the motor's eddy currents are dissipated across the shorted motor winding, and the wheel will almost lock up. I don't recommend shorting motor phases because the heat buildup in the motor coils may melt them.

I tried this, and it seemed okay with an X503 motor, but you do it at your own risk. The problem with this kind of regen is that the braking force is proportional to speed. At less than 5 mph, you have almost no braking force, but at 25 mph or more, the force is too strong. To reduce the braking force, you can add a resistor. Light bulbs make good resistors. The lower the resistance, the more severe is the braking. With light bulbs, the higher the power of the bulb, the higher is the braking force. The bulbs dissipate heat well and can be used as a brake light. Don't brake from too high a speed, or the voltage will be too much for the bulb to cope with, and it will blow. At different speeds, you need different resistances to have effective and safe breaking. A heavy-duty variable resistor would work, but it would be too bulky and unsafe to operate while riding. Possible solutions include a pic micro controller circuit, the use of two different resistors controlled by different buttons, or saving the brake for emergency braking situations only. The circuit diagram is shown in Figure 6.37 for brushed motors and in Figure 6.38 for brushless motors.

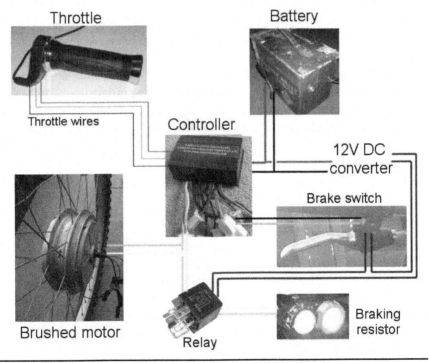

**FIGURE 6.37**  Plug braking for a brushed motor. Two brake switches are used. One tells the controller to switch off, and one activates the relay for plug braking. This is to avoid accidentally shorting the motor while the throttle is engaged, which would blow the controller.

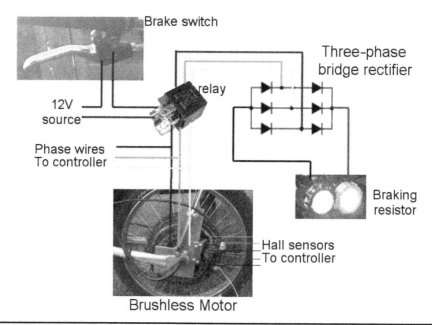

Brake switch

Three-phase
bridge rectifier

12V
source

relay

Phase wires
To controller

Braking
resistor

Hall sensors
To controller

Brushless Motor

**Figure 6.38**  Plug braking for a brushless motor. For brushless motors, there are some changes needed for the previous design. A three-phase rectifier is needed to include all motor windings equally in the braking effort. The same double-brake-switch design should be used for protecting the controller.

1.  Make a coil of insulated wire of about half an ohm, or use two 75-W 12-V light bulbs in series. This will be the braking resistor, and it should be located somewhere with good airflow for plenty of heat dissipation.

2.  Wire the braking resistor up to a DPDT relay, and connect it to the two outputs from the motor. The relay should be capable of withstanding 30 A. The connection to the motor controller should be made using a junction box so that the relay can be removed easily if it melts.

3.  Use a brake-mounted switch to control the relay. There are purpose-build ones sold on eBay, or you can make your own with a reed switch and a magnet taped to the stock of your brake. The relay will need 12 V (usually) to switch, so the brake switch should be connected in series with the 12-V source and the relay, as shown in Figure 6.37. The 12-V source can be a dc converter (recommended), or you could tap your battery halfway through to create a 12-V source (not recommended).

4.  To prevent the controller-popping situation of shorting the controller output while using throttle, another brake switch is needed. The brake switches need to be installed on the same brake so that they activate together or the controller off activates first. This is a safeguard to protect the controller. You should never brake while using the throttle anyway, even in normal operation, because it is bad for the controller.

5.  For brushless motors, the three motor phases cause an extra headache. A three-phase bridge rectifier is needed to reduce them to two wires so that braking is smooth (see Figure 6.38). The alternative is to use three braking resistors between each of the three phases.

### Problems with DIY Regen Braking

To do something useful with this braking energy instead of just wasting it across a resistor requires a little bit more work. The voltage the motor produces during braking varies wildly and continuously depending on speed. The voltage can be anywhere between the full voltage of the battery, at top speed, and nothing, when stationary. Therefore, to use the voltage to charge the battery, you need a power converter that can step up the voltage and will accept a wide range of input voltages. You also need to limit the current so that it doesn't harm the battery. The regen current is what determines the braking force, so this maximum current limit also will determine the maximum braking force. The power requirements of the converter would be quite large, so it might be quite bulky.

There are other regen braking modifications that use two controllers, one to control normal vehicle speed and one to control regen braking, but this is too complicated. The best option is to use a single controller with a regen braking function. Some regen braking controllers have solved the problem of varying braking force by using the throttle in conjunction with a regen switch to let the rider set the amount of regen to be used. The vetrix electric motorcycle has a special hand throttle that you can twist in the opposite direction for regen braking. This is the solution that should be included on all electric bicycle controllers in the future. A good DIY solution would be to use a boost converter with pic micro controller.

### Moding the Infineon Controller for Regen

The Infineon controller supports regen braking, but it is not connected as standard. You have to open up the controller and connect a handlebar-mounted switch across BK and GND. When the switch is pressed, it will activate regen braking as long as the throttle is in the off position. If you put a jumper across BK and GND, then the controller will mimic engine braking.

### 6.2.2 Torque Arms and Disc Brake Mounts

If you are going to use regen braking of any kind, then you will need to use torque arms or the bicycle frame will gradually get eaten away by the constant seesawing of the axle in the dropouts. If the motor is a front hub motor in aluminum forks, this could fail spectacularly and send you flying over the handlebars. What is more likely, however, is that the axle would eat a larger hole in the dropouts, and this would cause the axle nuts to slip, making the wheel rub against the frame. The axle also could fail by spinning around inside the dropouts, which would damage or wreck the motor and maybe the controller too. Torque arms also should be used on aluminum frames with powerful hub motors in front hubs.

A torque arm works by extending the thickness of the 10-mm flat section that supports the axle. However, it is not just the 10-mm flats of the dropouts that provide torque support. The side tension from the axle nuts and the surface area that is under tension also add torque support. This is why hub motors have fine threads so that you can tighten the bolts up with loads of tension. The axle bolts never should be left with spare thread exposed, not all the way on the axle, or you will strip the thread if you tension the bolts with proper force. Buy a big socket wrench especially for tightening axle nuts. A socket wrench spreads the force onto all sides of the nut, whereas an adjustable wrench will just round the corners off the nut.

Some commercial torque arms are beginning to become available from electric bicycle retailers, but they are still a niche product and difficult to get. The simplest torque arm is just a 10-mm spanner across the flats of the axle (Figure 6.39). This can be cable-tied to the frame or rear rack or arc welded in place. Another method is to buy small off-cuts of steel and make a custom

FIGURE 6.39   A 10-mm spanner as a simple torque arm. This will help with powerful motors on aluminium dropouts, but it will not prevent dropout failure if regen is used even on steel frames. It would need to be solidly fixed to prevent torque movement. Welding is the best method.

torque plate. The steel should be at least 4 mm thick if it is to withstand re-generative braking. However, this is quite difficult for the do-it-yourselfer because steel is tough to drill and file. It is possible to send design drawings to companies over the Internet who will CNC cut the piece for you. An easier approach is to use thicker (10 mm) quality aluminum tool plate. This can be drilled and tapped while in position on the bicycle. The advantage of a custom torque plate is that it can be used to add disc brake mounts to a non–disc brake frame.

I have used this technique successfully to convert several bicycles to rear disc brakes where there were no disc brake mounts. A 10-mm aluminum torque plate, when used with steel frame dropouts, may be enough to prevent dropout erosion caused by regen braking. If you can arc weld, an easier method would be to just weld some chunky metal to supplement the 10-mm axle flats.

Disc brake mounts can be welded onto frames too. To weld disc brake mounts, you have to bolt the disc brake onto the spare mount and position it on the disc rotor. Pull the brakes on to line up the disc brake mounts into the position to be welded. Then tack weld them on. Finish the job properly without the disc brake in place so that the welding doesn't melt the brake.

### Making a Custom Torque Plate

A custom torque plate can be made without any specialist machining but is time-consuming to build this way. The torque plate is designed to go on the axle inside the dropouts and provide more area for gripping the axle.

1. Buy a cut of aluminum 200 × 100 × 10 mm thick.
2. Line it up with the dropout, and mark where you need to cut it. Mark out where the axle is, including the angle of the axle flats. Mark out where the cutout sections for the chain stay and dropouts will be (Figure 6.40). If you are using this plate to support a disc brake, then line up the disc brake caliper to see the angle at which it needs to clamp the disc and the angle the support hole in the plate needs to be.
3. Drill the hole for the axle and the hole for the motor wires to come out. The hole for the axle will be a 10-mm-wide rectangle. You can make this by drilling two holes and then filing out between them. You also will need a hole for the wiring to emerge from the motor. The angle on the axle flats is critical, which is why you do this first and then build everything around it. Center punch the holes before you drill, and don't overlap the holes or the drill will slip into the first hole when drilling the second hole. Be patient, and don't rush. This is much easier and more accurate if done on a proper milling machine. File the edges smooth so that they don't cut the wires.

**FIGURE 6.40**    Lining up the torque plate and marking out the sections to be cut out for the chain stay and dropouts. The hole in the middle is for the axle and the exiting wires.

4.  Put the axle hole over the axle, and pull the motor wires through the wiring hole. Check that the sections marked out for the dropouts and chain stay are still in the right place. Make any corrections necessary.

5.  Cut out the sections for the dropouts and chain stay. Use a hacksaw and a drill. Cut two parallel cuts into the plate, and then terminate these cuts with drilled holes. To join the cuts up, sideways saw using the hacksaw, and the section will be cut out. File the edges smooth.

6.  Now you can put the torque plate on the bicycle. It should go flat against the flat metal section of the dropout. It then will extend toward and go around the chain stay, supporting the torque from there. Putting the motor on the bicycle with the torque plate is usually pretty difficult, but it will fit with a bit of encouragement. Use cobra half tubes for easy puncture repair.

7.  When it is all secured in position, drill holes for the rack mount and disc brake caliper and any other holes that you want that will add torque support or function. For the rack mount, you can drill with a 5-mm drill and then tap with a 6-mm tap. You don't even have to take it off the bicycle, which is a good thing, because it is difficult to do.

8.  You might need to use washers or axle spacers for clearance between the disc and the plate. You may need to make special C-shaped washers to slide onto the axle from the side if there is insufficient space to install

normal washers. This is easy to do just by sawing a chunk out of a normal washer.

9.  Remember to tighten the axle nuts hard. The result should look something like Figure 6.41.

**Figure 6.41**   The finished item. This torque plate provides disc brake caliper mounts as well as support for motor torque and a little extra support for the battery rack.

### 6.2.3  Preventing Sparking When Plugging In the Controller

It's a good idea to do this modification because it will mean that your controller and battery plugs will last longer. It's simple, using a precharge resistor to slowly charge up the controller capacitors prior to plugging in (Figure 6.42).

### 6.2.4  High-Powered Lights

Normal bicycle lights are usually only 0.5 to 1 W, and drivers complain that they cannot see them at night. By comparison, car head lamps are 55 W each, and there are four of them. Bicycle lights are limited by their little batteries and the desire not to have to charge them every day. Bicycle light technology has moved toward efficient light-emitting diode (LED) lights and flash-

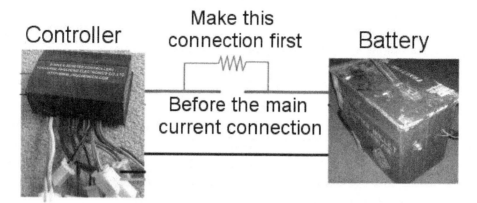

**FIGURE 6.42**    To prevent sparking on connection, plug in the negative wire as usual (black), and then make the positive connection with the small wire with the precharge resistor. This will slowly charge the capacitors in the controller. Then plug in the main positive connector, and there will be no sparks. Finally, unplug the precharge resistor.

ing lights to save energy and still be seen. In electric bicycles, however, the batteries are often bigger than car batteries and are charged every day, so their lights need not be limited in the same way.

The best, cheapest, and most readily available high-powered lights on the market are low-voltage halogen MR16s and MR11s. These operate on 12 V, come in different power options, and have mirrored backplates for focusing. Commonly available choices for MR16 are 10-, 20-, and 50-W bulbs, although 15-, 35-, 75-, and even 250-W bulbs are available. MR16s are 45 mm in diameter, but unfortunately, they don't have any holding mechanism because they are designed for ceiling spot lights. The bulbs can be overvolted to 15 V, and to meet the higher voltage of an electric bicycle, you can string them together in series. Series-connecting bulbs are good in this application because if one bulb is disconnected, all the bulbs in the same string will go out. If you series-connect the front and rear lights, you will know if the rear light has malfunctioned by looking at the front. This ensures that you will never unknowingly be driving without a rear light.

MR16s are halogen bulbs that are bright and reasonably efficient. A 10-W bulb front and back would be sufficient. A 48-V bicycle will need four bulbs, so 20-W bulbs would be overkill. As a side note, if your bicycle lights go dim when you accelerate, then your motor is putting too much stress on your batteries. Plastic pipe fittings make good bulb housings. If you use a 72-V system and you do not want to series-connect six bulbs, then you will have to use a dc convertor. Dc converter chips are available and are very cheap on eBay. They are about the size of a matchbox, so they can be positioned easily around the handlebars.

There are several other DIY bicycle lighting options available, including compact florescent light bulbs (CFLs) and LEDs. These technologies are about four times more efficient than halogen but about ten times more expensive for the same light output. Both CFLs and LEDs are available in 12-V options, so they can be used with a 12-V dc converter from an electric bicycle battery. LEDs can be series-connected to make up the voltage to electric bicycle battery voltage just as with the halogens. LEDs are available in an MR16 package, so they could be used in the same plastic pipe fitting as the halogen bulbs. Main-circuit-voltage CFLs are much cheaper and sometimes free because they are usually subsidized by government or electric companies. To operate main-circuit-voltage CFLs, you can use a 12-V dc converter and a small autoinverter to make main circuit voltage from the electric bicycle battery.

*Building Powerful Electric Bicycle Lights*
1. Go to your local hardware store and buy some 2-in elbow plastic pipe fittings. One elbow will make two light housings.
2. Saw the elbow in half.
3. Take one of the two halves, unscrew it, sandwich the bulb in between, and then screw it back together again. This is one light (Figure 6.43).
4. There are several ways to connect the bulbs. You can either use fittings or solder them. Soldering them is more reliable for bicycle lighting applications. There are also purpose-made lamp fittings available, or a 3-A terminal block also will work with a bit of trimming. To solder into the bulbs, you need leaded solder and a high-wattage soldering iron.

**FIGURE 6.43**   Bicycle lights made from MR16 bulbs and 2-in plastic pipe fittings.

5. Duct-tape the back of the light fittings to protect the lamp connections and make the light more reliable.

6. To secure the lamp holders onto the bicycle, you can drill holes in the pipe and either cable-tie or bolt the holder on. Good mounting points are on the rear rack or using the reflector-holding brackets.

7. For the rear light, you need a red screen. You can use cut-up old bicycle reflectors or any red plastic film; even red carrier bag material will work. Cut the red material to 45 mm diameter, and place it in front of the bulb; then screw it in place with the plastic cover nut.

8. Painting the light holders is difficult because spray paint just comes off; it's best just to leave them the natural color of the plastic pipe.

9. Use only 10- or 20-W bulbs in these because 50-W bulbs will melt the plastic if left on.

10. Buy a decent-sized switch from the component supply store, and solder wires onto it.

11. Hot glue and cable-tie the switch onto the handlebars of the bicycle in a convenient position for use while riding.

12. Solder wires from the front lights to the switch to the rear lights and then to a fuse holder and a plug (Figure 6.44). The plug should connect into the junction box made earlier.

13. The fuse should be just above the ampere rating of the lights. In a short circuit, it should blow before the main battery fuse blows.

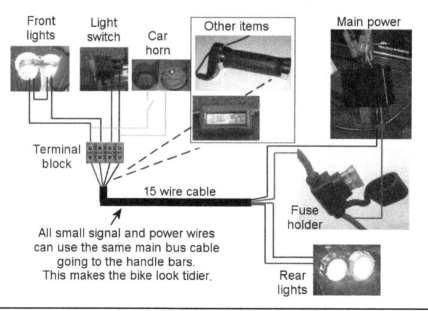

**FIGURE 6.44** Connection scheme for lights and auxiliaries with single cable to the handlebars. Front and rear lights in series to reach 48-V battery voltage. Battery voltage is available at the handlebars and fused for protection.

### 6.2.5  Powerful Horns

Have you ever been honked at by a car driver for no reason and not been able to respond because you were on a bicycle? Frustrating, isn't it? Well, now you can honk back louder! Electric bicycles operate on higher voltages than cars do, so you can have even more powerful horns than trucks. Two-tone car horns are available on eBay. They are usually colored red/black and come in a set of two, one for each tone. Each horn is 12 V, and you can put them in series for 24 V. However, since they are not used continuously, it's safe to overvolt them significantly. The same two-tone horn set can be used on a 48-V battery fully charged at 60 V. The horns come with a bracket for mounting, and they can be cable-tied to the frame using the holding bracket.

1. Bolt the horns together using a piece of angle iron or steel bracket from the hardware store.
2. Cable-tie the horns onto the bicycle where you want them.
3. Solder wires connecting the horns to the battery via a switch, as shown in Figure 6.44.
4. Hot-glue and cable-tie the switch to the handlebars in a convenient place that is accessible while riding.

### 6.2.6  Bicycle Mirrors and Turn Indicators

#### Mirrors

Bicycle mirrors are a very personal choice, and everyone has their own favorites. A lot depends on what sort of bicycle you are riding, what sort of helmet you have, and how the mirrors are adjusted. None of the commercial bicycle mirrors I have seen have been ideal. The helmet-mounted mirrors tend to vibrate and get misaligned when you put the helmet down. The handlebar-mounted mirror's view is usually blocked by your arms. The best handlebar-mounted mirrors are those on long stalks because you can see over the tops of your arms. These are designed for motorcycles but can be fitted to bicycles. Helmet- or sunglass-mounted mirrors have the advantage of allowing the user to move the field of view easily to see more stuff. There are some good "how tos" on the Instructables Web site at www.instructables.com regarding how to make a helmet-mounted mirror from an inspection mirror.

There may be an advantage to the inspection mirror approach over commercial designs because inspection mirrors are made of stronger stuff than plastic. If you buy the commercial design, you may end up having to modify it to fit your helmet anyway. With any head-mounted mirror, the helmet or glasses will need to be a really good fit to your head or it will bounce around too much. Even with a mirror installed, it is often quicker to look back than to check a mirror, and it's certainly safer because a mirror can have blind spots.

### Turn Indicators

None of the bicycle-mounted turn indicator boxes I have seen have been very good. The lights are usually too weak and too close together to be seen by other drivers. Often the lights come as a package with indicators, rear light, and stop light. The indicator switch is a single three-position slide switch that is mounted on the left handlebar. This is fiddly to operate because you have to move your hands from the handlebars and counterintuitive because it uses down for left and up for right. Some of the indicator kits have an annoying bleeping sound so that you don't leave them on. The brake light is activated from a clever switch mounted on the brake cable. There are other products that fit into the handlebar ends and have mirrors and indicators (e.g., Winkku). The best commercial indicators may be the wrist-mounted level-sensing type (e.g., Safe-turn). These are lights that automatically flash when you move your hand horizontally to make a turn. This is a simple system that allows drivers to see your hand signal at night. The bulbs look quite dim, though.

You could build your own turn indicators based on the preceding high-powered lights idea. This would allow you to custom build the indicators to your own ideals. A flasher circuit from Radio Shack could be used to make them flash. The front and rear bulbs could be put in the same pod using a 2-in plastic elbow as the housing or 1-in pipe and MR11s for a more compact pod size. Switches could be mounted on both handlebars so that you don't have to move your hands, and they could be spring-loaded switches to ensure that you don't leave them on by accident. If left on by accident, the 20-W bulbs would be obvious anyway, and drivers couldn't miss them. The pods could be mounted on the handlebar ends or cable-tied to the most outward part of pannier bags.

### 6.2.7 Heated Handlebars

Heated handlebars will be really useful to those who ride in cooler climates. Often in subzero temperatures your hands will freeze up and become desperately cold after only a short time of riding. It doesn't matter what sort of gloves you are using because the airflow will suck the heat out of them, especially if they become wet. This can be dangerous if you have to ride like this, so I have put this modification under safety modifications. Heated handlebars are an easy modification because you already have a convenient handlebar power supply from the lights and horn modifications. There are heated gloves on the market, but they have really weak heating elements.

They run on small batteries that are depleted in 1 hour of continuous use. The commercially available  gloves are low quality, the heating elements are in the wrong place for riding, and you can barely feel the warming effect. This handlebar modification will give you strong heating to the inside of your gloved hands, which will keep them toasty and warm.

1. Work out what resistance nichrome wire you will need for your battery voltage using Ohm's law (see Section 6.1.1, Equation 6.2) and based on needing about 5 W and 0.5 m/2 ft per grip. For example, for a 48-V battery, you will want to use 200 $\Omega$/m or 66 $\Omega$/ft wire; for 36 V, use 100 $\Omega$/m or 33 $\Omega$/ft; and for 24 V, use 50 $\Omega$/m or 17 $\Omega$/ft. You will need 2 to 3 m/6 to 10 ft of the wire for one set of handlebars.
2. Wrap 1 m/3 ft of the wire around each rubber handle grip. Make sure that the loops don't overlap, and then tape the wire loosely in place.
3. Temporarily connect the handle grip wires together and to the battery with crocodile clips to test the strength of the heating. If the heat is too little, then take off some of the wire and retry. If it is too much heat, then start again with larger pieces. Keep the wire the same length on each handle grip or the heating will be uneven. The hand throttle will be difficult to do because it needs to move. It is best to wrap wire only around the handle bit and not the throttle bit.
4. Once you find the right level of heating, then either duct-tape or plastic-tape the wire in place to make your handle grip heaters.
5. Solder wires onto a switch, and then hot-glue and cable-tie the switch onto the handlebars somewhere out of sight.
6. You can't solder onto nichrome wire. Therefore, to permanently connect up the handle grip heaters, you have to just twist them together with copper wire and then solder the copper wire. The solder will adhere only to the copper wire, but it will encapsulate the nichrome wire around which it is twisted.
7. Tape up the joins with electrical tape.
8. Connect the heaters in series and to the switch and then to the battery voltage at the handlebars.
9. To make sure that you don't leave the heaters on and drain the battery, you should install an LED indicator (Figure 6.45). This is just a red LED and resistor in series that then go in parallel with the power to the heaters. The resistor value needed for a red LED is 3,000$\Omega$ for a 48-V battery, 2,000$\Omega$ for a 36-V battery, and 1,200$\Omega$ for 24-V battery. The round bit of the LED should point toward the positive.

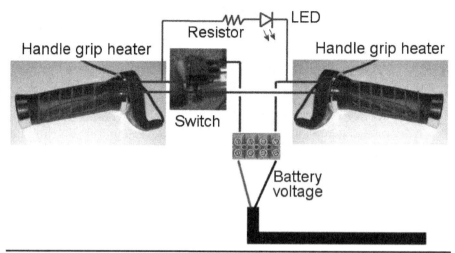

**FIGURE 6.45**   Handle grip heaters with indicator LED light to prevent accidentally draining the battery.

## 6.3  Other Modifications

### 6.3.1  Projects for Used Electric Bicycle Batteries

An average electric bicycle should use up about one battery pack every 3 or 4 years if the pack is treated correctly. However, if the batteries are mistreated by the user, then battery turnover will be higher. For this reason, it is best to learn and make your mistakes on a cheaper battery chemistry or read this book and not make those newbie errors. What do you do with all those reduced-capacity batteries? Well, thankfully, there are many projects that use batteries, and in time, I imagine that they will develop a resale value on eBay. I am writing this chapter now on a laptop powered exclusively by recycled nickel cells. These cells were ruined by bad battery carriers and experiments with DIY chargers. The nickel cells are down to half capacity, but that still gives me 1 hour of laptop time. Other interesting uses for old batteries include renewable-energy storage to recharge your bicycle and turning your car into a DIY hybrid vehicle.

**Laptop Battery from Recycled Electric Bicycle Nickel Cells**
Laptop batteries are notorious for dying after a year of use and are expensive to replace. Fortunately, there is a cheaper solution. You can recycle and repackage old cells from your electric bicycle. Most laptops run on 12 to 28 V and draw a power of about 4 A. Old electric bicycle cells with capacity reduced to 4 Ah will give 1 hour of laptop time, or you can parallel cells to in-

crease the run time. The drawback is that the recycled battery will be heavier than a replacement battery. Assuming that your laptop still works running off the wall charger, then you can build a battery pack for it.

1. Disassemble the old batteries and test the cells to find the good ones. Only use cells with voltages that are within safe operating limits for that chemistry. For nickel cells, this is 0.90 to 1.44 V; for $LiFePO_4$ cells, this is 2.5 to 3.7 V.
2. Decide how many cells you will need and your charging strategy. For my laptop, I have a 19.4-V charger, so I use just enough cells to be fully charged at 19.4 V. In this way, I know that I will not exceed the laptop voltage requirement, but I will have the right voltage to charge the cells. 19.4 V/1.44 V = 13.47. I rounded up to 14 cells.
3. Connect the cells to make a battery pack. Follow the battery pack building guide in Section 6.1.1.
4. Cut the dc side plug off the charger, and use it for your battery so that you can plug your battery into the laptop where the charger used to go. Make sure that you polarity check the connections so that you don't reverse the voltage.
5. Buy a mating plug to fit to the charger so that you can charge your battery. Check that the polarities are correct.

### Renewable-Energy Storage from Old Electric Bicycle Cells

Once you have a battery-powered vehicle, for maximum bragging rites, you may feel the desire to try to use renewable energy to charge it. The problem is that a renewable-energy supply fluctuates and only works when it's windy or sunny. So you can't rely on it to be available when you want it. To get around this, you need to have energy storage. The best energy storage is cheap batteries. Usually people use cheap lead-acid batteries, but if you have old EV batteries for free, then this is even cheaper. For solar panels, you will need a charge controller to keep the solar cells at peak efficiency. For a wind turbine, you probably can get away with just a blocking diode. The following instructions show you how to build a solar power system and a wind energy system.

*How to Build a Solar Power Energy Storage System*
1. Parallel lots of cells to make a large-capacity 12-V battery out of all the old cells you have. The fully charged voltage should be the same as that of a lead-acid battery because that's what charger controllers are designed for (14.4 V). For nickel cells you would want 10 in series; for $LiFePO_4$ cells, you would want 4 in series.
2. Buy a solar panel charge controller, and connect it to the batteries and solar panel as shown in the instructions. The charge controller also should protect the battery bank from overcharging.

3. Buy an inverter, and connect it to the 12-V battery.
4. Connect your bicycle charger to the inverter, and charge your bicycle as normal through the inverter.

*How to Build a Wind Power System*

1. Decide on a system voltage, and parallel lots of cells to make a large-capacity battery of that voltage: 12, 24, 36, 48, 110, and 240 V are popular choices. Chargers and power supplies work equally well on dc and ac power. It's best to still use your bicycle charger to control charging of your battery. Therefore, you can either use a lower-voltage system with an inverter to run the charger or you can have a higher-voltage system to directly run the charger. Some bicycle chargers may not need an inverter when operating from low voltages. It all depends on what the charger's cutoff voltage is. Some will take power from as low as 40 V. Most chargers have a wide input specification, usually 90 to 130 V for a 110-V supply and 220 to 250 V for a 240-V supply.
2. Figure out how to generate that voltage. Electric drills or old electric bicycle motors make good wind turbines, or use solar panels.
3. Connect the generator via a nonreturn diode to the battery pack. This is done so that the batteries don't push power back to the turbine when it is not generating.
4. Sort out the charge-termination method so that the battery bank doesn't get overcharged. For nickel chemistries, you want a thermostat-operated system; otherwise, you could use a high-voltage cutoff circuit. The charger's cutoff voltage will protect the battery bank from being overdischarged by your bicycle, or you could use a low-voltage cutoff circuit.

## 6.3.2 Renewable-Energy-Powered Electric Bicycles

It is impractical to expect people to power a whole car by renewable energy; the costs involved are beyond the means of most people. An electric bicycle, however, uses much less energy and is the perfect candidate for renewable-energy-powered transportation that is well within the means of many people. For the environmentalists out there, powering your electric bicycle with renewable energy is less important than simply using the bicycle to cut down on car use. There are much more $CO_2$ savings to be made from not using your car than from not using fossil fuel to power your bicycle. Many people, however, will find that the bragging rights for renewable-energy-powered transportation are priceless! The best renewable-energy source for electric bicycles is solar power. Solar is easier to have in residential areas because it makes no noise and doesn't require a tall, heavy thing that could blow over in high winds. Wind power, though, is cheaper and will work better in rural, non-built-up areas. Solar power could be made portable for an electric bicycle, either in an unfolding suitcase solar panel or a very small panel permanently attached to the bicycle. Wind turbines also could fulfill the power needs of an electric bicycle, although they are not as portable or as reliable as solar.

### Wind and Hydro

Old electric bicycle parts make great wind turbines. In fact, it's very easy to turn an electric bicycle into a wind turbine. If the bicycle has a regen braking–enabled controller, then you can simply use duct tape on the spokes to create diagonal facing "blades." Then with the electric wheel off the ground and the bicycle side-facing into the wind, this should be enough to generate some power. If the electric wheel is on the front, then the handlebars can be used to pivot the turbine much like a tail fin on a conventional wind turbine.

A similar thing could be done for hydroelectric power if you live near a fast-flowing stream. Dip the electric wheel just so the tire is beneath the water. The flow of water should turn the wheel, and the regen will be enough to generate power. Be careful not to fall in the water, though, or drop the bicycle in! For a more fit-for-purpose setup, the motors from electric bicycles make great wind turbines. Turbine blades can be cut easily from drain pipes and mounted onto a hub motor or chain-drive motor. Brushed motors would make dc electricity without power conversion; only a blocking diode would be needed to prevent discharge when there is no wind. Brushless motors would need only a three-phase bridge rectifier to convert the power output to dc (see Section 6.2.1). Old worn-out electric bicycle batteries work great for stationary renewable-energy applications (see Section 6.3.1). An alternative to battery storage is to use a grid-connected system. A grid-tied system allows you to export power back to the electricity grid when it's not needed. A grid-tie inverter allows you to feed power back into a standard house socket, which is very convenient but slightly expensive at present.

### Solar-Powered Bicycles and Chargers

Solar power is probably the best renewable source for electric bicycles because, unlike wind or hydro, it can be used in normal riding. In some parts of the world, near the equator, for example, where the sun is strong, this could be used to slightly extend electric bicycle battery range. A wind turbine fixed to an electric bicycle could not be used to charge the bicycle while riding (unless a tailwind is blowing you forward). This is so because the air drag on the turbine would more than cancel out the energy generated by it; otherwise, it would be perpetual motion, and that is not possible. One would have to be careful of the solar panels creating drag as well, but this is not how solar cells derive their energy, so it doesn't have to be a problem. The best photovoltaic (PV) cells on eBay now are priced at $2 a watt, so a 100-

W(peak) solar charger could be made for as little as £133/$200. The (output) energy density of these solar cells is 177w/m² or 16w/ft², which would make a 100-W(peak) solar charger 0.6 m²/6.3 ft² in size. The weight of this solar charger would be around 9 kg/20 lb. This is almost feasible to install on a standard bicycle, although it would take up a lot of room. It would be more feasible to install it on a trailer or trike.

Unfortunately, it is unrealistic to assume that the sun shines all the time and that you can always achieve the rated power of the panels; also solar cells are only 20% efficient. On average, in Britain, solar panels generate 22 W/m² or 2 W/ft². America is closer to the equator so more solar energy can get through the atmosphere. In New York average solar panels will generate 30w/m² and in LA 45w/m² (Mackay, 2009). This is a lot less than what the panels are capable of collecting, which makes significant range extension from onboard solar unfeasible. The only solar step-up that would make sense would be a larger stationary but maybe portable, foldable solar charger.

### Solar Cell Technology

There are three types of solar cells: amorphous, polycrystalline, and monochrystalline. Amorphous is the worst kind, and monocrystalline is the best. It is possible to buy the cells separately to build your own panels at a cheaper price or buy ready-made panels with a guarantee at a higher price. The performance of solar cells can be increased to their rated specs if you are living in low-light areas by using techniques such as concentrating lenses, mirrors, cooling, and sun tracking. Small solar concentrating systems can be very powerful and burn things in seconds. Check out www.greenpowerscience.com for more information and some truly inspiring youTube videos.

It is important to keep the cells cool if using concentrating mirrors or lenses because temperatures above 25°C/77°F reduce the cell's output, and high temperatures can damage them. Water or oil cooling therefore is important for cell performance and longevity. Performance drops by 0.5% for each degree above 25°C/77°F (Yates, 2003).

PV cells have an open-circuit voltage of 0.5 V. They are very fragile and need to be housed inside a sturdy box with a transparent lid for protection during operation. Solar cells can be broken in half to create two working cells with half the power output but the same voltage as the original cell. The positive and negative terminals are on either side of the glass surface. Broken and misshapen cells are available on eBay, but the price per watt is not significantly lower than that of regular cells.

There are several other solar technologies that show promise, such as concentrated solar steam engines and concentrated solar Stirling engines. However, none of these technologies seems to be available commercially at a competitive price. It is possible but time-consuming to build these solar engines at home. Small amorphous solar cells also can be made at home.

# Glossary of Terms

| | |
|---|---|
| ac | Form of electricity supplied by main circuit power |
| BMS | Battery management system |
| Bridge rectifier | A method of converting ac electricity to dc |
| CC | Constant current |
| C rate | Rate of charge or discharge relative to a battery's capacity; for example, 1C means that the capacity of the battery will be used/charged in an hour. A battery's current is related to its capacity by this number. |
| CV | Constant voltage |
| Continuity test | Checking a circuit for a through connection |
| dc | Form of electricity supplied by batteries |
| Discharge rate | How fast you can discharge a cell or battery without damage; measured in multiples of the capacity of the battery, for example, C rate or 1C, 5C. A battery's current is related to its capacity. |
| DPDT | Double-pole, double-throw; refers to relays and switches that can switch several things at once. |
| Dropouts | The part of the frame that holds the wheel axle; usually refers to the part of the frame holding the rear wheel. The front dropouts are called *forks*. |
| EV grin | The feeling you get when you first ride an electric bicycle; caused by the sensation of effortless movement and speed. |

241

| | |
|---|---|
| FET | Field-effect transistor |
| LCD | Liquid-crystal display |
| LED | Light-emitting diode |
| LiCo | Lithium-cobalt |
| LiFe | Lithium-iron |
| LiFePO$_4$ | Lithium-iron-phosphate |
| LiPo | Lithium-polymer |
| LiMn | Lithium-manganase |
| LVC | Low-voltage cutout; a safety feature on many controllers |
| NiCd | Nickel-cadmium |
| NiMh | Nickel-metal hydride |
| No-load speed | Top speed of a motor when it is spinning freely |
| OCV | Open-circuit voltage; the voltage of a battery when it is not being used |
| Open circuit | When there is no circuit made and current cannot flow |
| Opportunity charging | Charging whenever there is a socket available, for example, on a train station platform, on a train, etc. |
| Overclocking | Increasing the speed of some standard piece of kit past factory settings; usually at the risk of overheating it |
| PbA | Lead-acid battery |
| Plug braking | Using the motor for braking; in plug braking, the energy is wasted across a resistor and/or the coils of the motor. Similar to regen braking but without charging the battery; can produce very powerful braking force. |
| Polarity | The way in which the positive and negative wires are arranged |
| Pot | Potentiometer variable resistor |
| PV | Photovoltaic |
| PWM | Pulse-width modulation; varying the power of a motor by changing the time duration of the pulses during the revolution of the motor. The power and frequency of the pulse are the same, but its duration is varied to control the average power. |

| | |
|---|---|
| RC | Remote control (cars, airplanes, etc.) |
| Regen | Regenerative braking; generating power from waste kinetic energy when braking. The motor is used in reverse to generate electricity and charge the batteries using the kinetic energy of the vehicle and thus slowing it down. Related to plug braking, which is where the energy is not used to charge the batteries but wasted as heat. |
| Short circuit | When a load is bypassed and the power source may cause damage to itself |
| SLA | Sealed lead acid |
| Stealth | Stealth is concealing the fact that your bicycle is electric. This may be for many reasons: So that it looks like you are an athletic cyclist, to conceal its true value from thieves, or so that you don't get stopped by police if the laws in your country prohibit electric bikes. |
| Stinger | A small piece of wire connected to a meter probe or soldering iron that allows it to get into difficult-to-reach areas |
| Tap | A connection between cells or batteries that can be used to monitor battery voltage or provide a lower voltage |

# References

Buchmann, I. 2001. *Batteries in a Portable World: A Handbook on Rechargable Batteries for Nonengineers*, 2d ed. Cadex Electronics, 22000 Fraserwood Way, Richmond, BC, Canada V6W 1J6. Available at www.buchmann.ca/toc.asp.

www.confused.com car insurance price comparison Web site; accessed October 21, 2009. The lowest car insurance quote for a 17-year-old male driver was £4,500/$6750. The quote was for third-party fire and theft insurance on a Citeron AX 1.0L £1,000/$1500 car.

Green Power Science. YouTube videos involving Fresnel lens and parabolic mirror solar concentrating systems. Available at www.greenpowerscience .com. Accessed October 8, 2009.

Lemire-Elmore, J. 2004. The energy cost of electric and human powered bicycles. Available at http://ebikes.ca/sustainability/Ebike_Energy.pdf. Accessed September 14, 2009.

Mackay, D. J. C. 2009. Sustainable energy—without the hot air. UIT Cambridge, Ltd., pp. 118–140. Available at www.withouthotair.com.

Shinnar, R. 2003. The hydrogen economy, fuel cells, and electric cars. *Technology in Society* 25:455.

van Mierlo, J., Maggetto, G., and Lataire, Ph. 2006. Which energy source for road transport in the future? A comparison of battery, hybrid and fuel cell vehicles. *Energy Conversion and Management* 47:2748.

Yates, T. A. 2003. Solar cells in concentrating systems and their high temperature limitations. Bachelor's degree thesis, Department of Physics, University of California, p. 7.

# Index

CPSIA information can be obtained
at www.ICGtesting.com
Printed in the USA
BVOW04s0236101117
499866BV00052BA/777/P